Lotus *Illustrated* DICTIONARY *of* TEXTILE

Compiled and Edited by:
Larry Operah

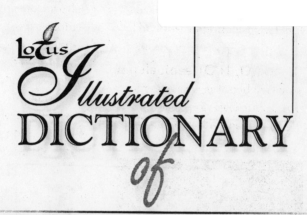

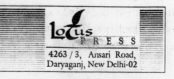

4263/3, Ansari Road,
Daryaganj, New Delhi-02

Lotus Illustrated DICTIONARY of TEXTILE

All right reserved. No part of this publication may be copied, reproduced, stored in a retrieval system or transmitted in any form or by any means without specific prior permission of the publisher.

© Lotus Press: 2010

ISBN 81 89093 62 2

Published by:
LOTUS PRESS
4263/3, Ansari Road, Daryaganj,
New Delhi-110002
Ph: 32903912, 23280047
E-mail: lotus_press@sify.com
www: lotuspress.co.in

Printed at: **Chetna Printers**, Delhi

PREFACE

'Clothes are the representation of a man's personality', a well-said quotation. No deny in the fact that they are considered to be amongst bare necessities of life. Clothing and the type of texture worn by a person depicts his or her personality, as stated above. The way this Global-world is moving, and the times are altering, there is a drastic revolution in the Textile field also. The fashions in this field have undergone different variations, as per the taste of the people. The quality of texture and the fashion in the outside world are improving day by day. With the quick and timely variations in this sector there is a deep impact on the choice of people. And now a days the future of fashion or the latest trends in the fashion doesn't last for a longer period.

This dictionary helps the reader to discover even the minutest terms about textile and fibre and it keeps the man up-to-date with the social and fashionable changes. The dictionary also helps the reader to become aware about the different variations in the fashion field. The terms are explained precisely in a manner that the reader finds it interesting to discover and further adds to his knowledge.

abaca

a kind of hemp like fibre used for cordage grown in the Philippine Islands. It is obtained from the plant Musa, used for Textile making, commonly known as Manila. This vegetable leaf fibre is derived from the Musa textiles plant. Also found in Africa, Malaysia, Indonesia and Costa Rica, in smaller quantity, it is obtained from outer layer of the leaf. Etching occurs when it is separated mechanically decorticated into lengths varying from 3 to 9 feet. Abaca is very sturdy, has great lustre and is resistant to salt. It is used to make Cordage.

abho

a loose shirt-like garment, worn by women mostly in Gujarat and Rajasthan. Engraved with embroidery and mirror-glass work, the garment was generally worn with short, wide sleeves, open at the neck, loose-fitting on the upper part and really radiated in its skirt.

abrasion

the wearing away of fibre by rubbing away.

absorbency

the ability of a fabric to consume in moisture. Absorbency is a very important property, which affects many other characteristics such as skin comfort, static build-up, shrinkage, stain removal, water repellence and wrinkle recovery.

absorbency under load

the weight of fluid in grams that can be absorbed by 1 gram of fibre, yarn or fabric which has been subject to a pressure of 0.25 lb/in^2 before wetting.

abstract

refers to a design in the epitome style, i.e. one that represents a conventional form ignoring a precise representation of a subject.

accessories

residual embellishment to the garment in order to create a miraculous appearance. They are used in shoes, jewelleries, garments etc.

acetate

a manufactured fibre named cellulose in which the fibre-forming substance is cellulose acetate. Acetate is a fibre that resists shrinkage, moths and mildew and is not good moisture absorbent. Its yarns are pliable and supple and always sprig back to their original shape. It is fast drying and becomes more pliable when heated.

acetone-soluble cellulose ethanoate (acetate)

hydrolysis of primary cellulose ethanoate (acetate) when allowed to precede combined ethanoic (acetic) the product is the remains from the acid. It's soluble in propane (acetone) and is also known as acetone-soluble cellulose acetate.

■ **acetylation**
process of introducing an ethanoyl (acetyl) radical into an organic molecule.

■ **achkan**
a long-sleeved coat-like garment, covering from neck to knees or even below and buttoned in front-middle.

■ **acid dyes**
being water soluble they are applied directly with an acid, such as sulphuric acid. Bright colours do not stand up too well in colourfastness when wet-treated, fair to poor in washing, good in dry cleaning and light-fastness. Used on wool, worsted, acrylics and nylon.

■ **acid-milling dyes**
ideal for colouring wool, worsted, acrylics and mod acrylics, nylon and spandex fibres. Also used in printing. Good in dry-cleaning, excellent resistance to light.

■ **acid-pre-metalised dyes**
used on wool, acrylics and nylon, they rate excellent to fastness in dry-cleaning, perspiration and washing. Much used for carpeting, suiting fabrics and upholstery.

■ **acrylic**
a manufactured fibre derived from polyacrylonitrile. Its major properties include a soft, wool-like hand, machine washable and dry-able and excellent colour retention. Solution-dyed versions have excellent resistance to sunlight and chlorine degradation.

■ **acrylic**
a kind of manufactured fibre of acrylonitrile. Though it has an uneven surface, but it's durable fibre with a soft, woolly feel. It can be found in different colours and is sun and chemical resistant. It can be used as a replacement for wool.

■ **acrylic coated**
a fabric coated with acrylic resin on the back, which makes it waterproof or dawn proof.

■ **adjective dye**
a dye, which requires the use of mordents.

■ **affinity**
the quantitative expression of substantiality. It is the difference between the chemical potential of the dye in its standard state in the fibre and the corresponding chemical potential in the dye bath.

■ **agneline**
a black woollen, crude and heavy fabric with a very long cervix. On stretching it becomes tightened and water-resistant.

■ **aguillettes**
metal-tagged laces to replace the woven ones and attach the pantaloons to the doublet.

■ **air laying**
a method in which fibres are first dispersed into an air stream and then condensed from the air

stream on to a permeable cage or conveyor to form a web or batt of staple fibres.

■ **air texturing**

a process in which yarns are overfed through a turbulent air stream so that entangled loops are formed in the filaments.

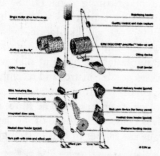

■ **air-laid**

a web or batt of staple fibres formed using the air laying process.

■ **ajouré**

an embroidery technique that creates open areas, often in figured patterns and usually on a woven fabric.

■ **akha spindle**

a lightweight, supported spindle.

■ **aklae**

norwegian low-warp tapestry technique. Wefts interlock between two warp ends.

■ **albatross**

a fabric with slumber, fleecy surface, light in weight, plain weave traditionally of wool. Its light in colour and is generally used in infant's wear or sleepwear.

■ **albert cloth**

a cloth with double layer of wool, which is curable and provides additional warmth to body. Its faces and backs vary in colour.

■ **alencon lace**

a design outlining made with needlepoint lace on a fine net ground classified by a heavy thread (cordon net) outlining the design. It's usually done with machine but at times it is inserted by hand.

■ **alginate (fibre) (generic name)**

a term used to describe fibres composed of metallic salts of alginic acid.

■ **alizarine dyes**

originally natural dyestuff, they are now synthetic dyes. Used on cotton and wool, they are resistant to sunlight and washing. Considerable use for apparel fabrics.

■ **alkali-cellulose**

an output to the interaction of strong sodium hydroxide with purified cellulose. In the manufacture process of viscose fibres, the cellulose can be cotton linters or wood pulp.

■ **alley**

the measure between the breaker carding and finisher carding machines in which the working of alley tender.

■ **alligator skin**

a printed or engraved design, which suggests the characteristic texture of an alligator.

■ **allonge-perruqe**
a French term for periwig, also known as state-wig. Fashionable men of late 17th and early 18th century wore it. It was extremely long and had very high 'horns' on top of the forehead. It curled and flowed down the back, over the shoulders.

■ **allover lace**
a term used for a wide lace, which covers the full width of the fabric. Similarly, like non-lace fabrics it is sold and cut.

■ **aloe**
a plant whose extract is believed to have a beneficial effect on skin.

■ **alpaca**
1. cloth of fine, silken nature, soft in feel, light in weight. The fibre is obtained from the animal of that name. The yarn is often used as filling in some cotton warp clothes. Alpaca resembles mohair and is imitated in cheaper cloths or those in combination with the genuine. The cloth has much lustre and is beardy in some instances. Much alpaca is now made from wool-and-rayon blends. It is used for women's spring or fall coats, suits and sportswear.
2. a product of the natural hair fibre of alpaca sheep. Its generally used in making of dresses, suits, coats and sweaters. It is also traced in wool, rayon, mohair, rayon or cotton. Alpaca filling also provides synthetics - e.g. orlon. Fine, soft, silk-like fabric, light in weight and warm. A strong and durable fabric, rich and silky with considerable lustres and resembles mohair. It can be found in white, black, fawn or grey. The fibres are less coarse than those of the llama but are higher in tensile strength.

■ **alter**
changing of the pattern to correspond it to body measurements.

■ **alum**
hydrated double-sulphate of alumina potassium. A commonly used mordant.

■ **amadis sleeve**
a tight-fitting sleeve perennial on the back of the hand.

■ **American pima cotton**
a cross between Sea Island and Egyptian cotton. Grown in Ari-

zona. Length averages 13/8" to 15/8".

■ **ammonia**

an alkaline liquid used in natural dyeing.

■ **anaphe**

a wild silk obtained from the larvae of the Anaphe moth.

■ **angarakfia**

a linear, full-sleeved outerwear for men, which protects the limbs. It is opened at the chest and fastened in front, with an inner flap or a 'parda' covering the chest. It can be found full-skirted and of varying lengths.

■ **angiaiangika**

a kind of covering for the body. Its short, tight-fitting corset worn by women of India from very early times.

■ **angora**

downy soft, fluffy hair that is plucked or sheared from the angora rabbits.

This is a slippery, flyaway fibre and is usually blended with wool or other fibres to make it easier to spin and to reduce the cost.

■ **angora goat**

the goat that produces 'mohair'.

■ **angora rabbit**

hair from the angora rabbit, blended and mixed with wool to obtain fancy or novelty effects. It is long, very fine, light weight, extremely warm and fluffy fibre and has a tendency to shed and mat with time. The fibre is used mostly in knitwear - gloves, scarves, sweaters, etc. for children and women. It gives a softer feel if unified with wool in dress goods and suits.

■ **anidex (fibre)**

a term used to describe fibres made from a synthetic linear polymer.

■ **aniline dyes**

a class of synthetic, organic dyes originally obtained from aniline (coal tars) and were the first synthetic dyes. Today the term is used with reference to any synthetic organic dyes and pigments, in contrast to animal or vegetable colouring materials and synthetic inorganic pigments. Aniline dyes are classified according to their degree of brightness or their light fastness. Also called 'coal tar dyes'.

animal fibres

these include alpaca, angora goat hair, camel hair, cashmere, cow hair, fur, guanaco, hog hair, llama, mohair, misti, Persian cashmere, rabbit hair, silk, spun silk, suri, vicuna, worsted, worsted top.

animal skin

a kind of design that gives an appearance of the skin of an animal. Some of the popular motifs are leopard, tiger, zebra and giraffe.

anionic dye

a dye when scattered in aqueous solution gives a negatively charged ion.

anisotropic

a material, which has different physical properties in different directions.

antheraea spp.

see **tussah silk**

anthrax

a highly dangerous, infectious disease cased by Bacillus anthraces. In humans, a form of this disease is commonly called 'wool sorter's disease'. It may be contracted, most likely through skin abrasion from handling fleeces from infected animals.

anti bacterial

a finish that makes a fabric bacteria-resistant.

anti pill

a finish applied to deceive which involves the process of trimming the surface.

anti-dumping duty

an extra duty imposed on an imported product by an importing country (or group of countries, as in the case of the EU) to compensate for the dumping of goods by a foreign supplier.

antique satin

a reversible fabric, which has a dark warp, which enhances the texture. It appears like satin from one side and the shantung from the other. It is often used for draperies.

antique taffeta

a solid plain prismatic weave fabric, with a subbed weft. Could be of silk or synthetics.

AOX

Absorbable Organic Halogens.

apeo

alkylphenolethoxilate.

apparel wool

all wools that are manufactured into cloth for use as clothing.

apparent wall thickness

presumable width of a fibre wall as seen under the microscope. The apparent wall thickness is assessed visually on discretion test for cotton. Widest part of the fibres is visualised as a fraction of the maximum ribbon width.

application printing

also called commercial printing, it is actually a direct printing method. It is the printing of white goods irrespective of the type of goods involved, popular for cotton, rayon and some silk. No discharge or resist printing procedures are involved in this method of colouring. Usually applied only to low-grade materials where costs are involved so that the goods may sell 'for a price'.

■ applique

decoration in which material is cut out and sewed, embroidered or pasted on another material. Appliqué is used on lace, fabric or leather goods.

■ aramid

a strong fibre, which does not have a heating point and is flame proof. It is stretch-resistant and retains its shape, even at high temperatures. The generic name for a special group of synthetic fibres (aromatic polyamide) having high strength, examples are 'Kevlar' and 'Twaron'.

■ argentan lace

similar to alencon lace it's a needlepoint lace on a net ground. But it's done on a larger net and without the cordon net outline thread of alencon.

■ argyle

a diamond-shaped designing done on a single coloured ground, of multi coloured diamond shaped blocks or crossed lines. It's popular among sweaters and hosiery.

■ Arizona Egyptian cotton

cotton obtained from modified forms of Gossypium bard dense and raised in Arizona, New Mexico and Southern California. Staple ranges from 35 mm. to 45

mm. and includes Pima cotton raised in the area. Other cottons in this group include SxP, Amsak, Pima 32, Pima S-1 and Supima, the latter a trademark name.

■ armure

a kind of designing of two colours done on cotton, silk, wool, rayon, synthetics and blends. It has a plain, twill or rib, background often has a small design either jacquard or dobby made with warp floats on surface giving a raised effect.

■ arran

a traditional style of fishermen's cable-knit sweaters.

■ art linen

design woven with even threads that are especially good for embroidery and is done on linen. Threads are bleached or coloured and have a soft finish. It is used in all kinds of needlework, lunch cloths, serviettes, etc.

■ art/embroidery linen

a balanced plain weave fabric, made from smooth round yarns, usually of linen or linen/cotton. Also known as embroidery crash, it can be used as a base for embroidered table linen, pillowcases, in drapes, slipcovers and some apparels.

■ artificial wool

broad term for any fibrous material made to simulate natural wool.

■ asbestos

a universal term, which describes a family of naturally occurring fibrous, hydrated silicates parted on the basis of mineralogical features into serpentines and amphiboles. In fibrous form varieties of asbestos are deemed. But because of health risk it is no longer in use.

ASEAN
Association of South East Asian Nations-Brunei, Indonesia, Malaysia, Myanmar, Philippines, Singapore, Thailand and Vietnam.

asharfi buti
a famous textile design, comprising of small floral discs, having small patterns within the circles.

asphalt retention (geotextiles)
a measure of the amount of asphalt cement that can be held within the pores of a paving geotextile.

astrakhan
1. a thick woven or knitted cloth with a surface of loops or curls which imitates the coat of an Astrakhan lamb.
2. also known poodle cloth. It is a kind of thick woven or knitted cloth often of wool with a surface of loops or curls, repeating the coat of an astrakhan lamb. It is used in the making of coats and trimming.

atactic
a type of polymer molecule in which groups of atoms are arranged randomly above and below the backbone chain of atoms, when the latter are arranged all in one plane.

atactic polymer
a lineal polymer that contains disproportionate-substituted carbon atoms in the repeating unit of the main chain. Its planar projection structure has got same substituents situated casually to any one side or the other of the main chain.

atansaw
a wide garment used to wrap around the body, a kind of commodious 'chogha'.

ATC
the Agreement on Textiles and Clothing, which embodied the results of the negotiations on textiles and clothing conducted under the Uruguay Round of multilateral trade talks.

atlas
a drift fabric in whom a set of yarns shifts moves diagonally and then returns to the original position.

atmosphere for testing
1. standard temperature: an atmosphere at the prevailing barometric pressure with a relative humidity of 65% and a temperature of 20°c.
2. standard tropical atmosphere: an atmosphere at the prevailing barometric pressure with a relative humidity of 65% and a temperature of 27°c.

attenuation
in spinning, the fibres are pulled out of a distaff or from a ball of roving into a strand of the desired diameter.

■ **awning stripe**
heavy firm-woven cotton duck or canvas with either yarn-dyed, printed or painted stripes. Used for awnings, beach umbrellas, etc. Drills are used for inexpensive painted awning stripe fabrics.

■ **axle**
this is the metal shaft through the centre of wheel, supporting it. There is usually a setscrew that 'locks' it in place which (sadly) can be sheared off.

■ **azlon (fibre)**
a term describing the process of manufacturing, in which fibre-forming substance is placid of any regenerated naturally occurring protein. The ISO is the generic name of protein.

■ **azoic or naphthol dyes**
also called Ice colours, Insoluble Azos and Ingrain colours. Rated good to laundering and washing. Ideal for decorative fabrics, draperies, dress goods and sportswear.

■ **baby alpacer**
see **alpaca**

■ **baby combing wool**
short, fine wool, which is usually manufactured on the French system of worsted manufacture. This term is synonymous with 'French Combing Wool'.

■ **back**
the foundation of the cloth as woven in the loom.

■ **back beam**
see **breast beam**.

■ **back frame**
the side of a fly frame on which the bobbins, from which roving is drawn into the machine, are held.

■ **back rest**
see **breast beam**.

■ **backcross**
the mating of a crossbred animal to one of the parental breeds.

■ **back tanning**
posterior-treatment done by use of natural or synthetic tanning agents, to improve the wet fastness of dyed or printed silk or nylon, using either.

■ **badla**
silver-gilt, flat metallic wire utilised in brocading and embroidery.

■ **baghal bandi**
similar to bathrobe or jacket, worn with shorts and fastened under the armpits.

■ **bags**
in the United States, the commercial woolgrowers have their fleeces loaded into large cloth bags for

shipping to the wool mills. In Australia and New Zealand, the fleeces are packed into 'bales', which load better in the ship holds for export abroad.

■ **balabar**

an outer garment like 'achcan' or coat, worn by men.

■ **balagny cloak**

a clothe or cape with wide collar, popular in first half of 17th century.

■ **balanced**

a plied yarn that doesn't twist back on itself. If you hold 10 inches of yarn by the ends, then slowly move your hands closer together until they are 2 inches apart, a balanced yarn will drape itself into an elongated U. An over-spun yarn will ply back on itself.

■ **balanced stripes**

designing done by evenly placing of stripes in width and spacing.

■ **baldric**

a kind of external attachment like sword hanger usually decorated with exquisite embroidery (often metal thread embroidery) and is worn from the right shoulder to the left hip. Usually over the waistcoat or resembling earlier bolero-style doublet. It is worn under the coat.

■ **bale breaker**

a machine used for sorting cotton straight from a bale. In the process, withdrawn of layers of compressed cotton from a bale takes place and fed into a machine where the tearing action of two coarse spiked rollers moving in opposite directions, manufactures a more open mass of tufts.

■ **bale dyeing**

a low cost method to dye cotton clothes. The material is sent, without scouring or singeing, through a cold-water bath where the sized warp yarn has affinity for the dye. With the natural wax not removed from the filling, this filling will not absorb yarn the dye. Imitation chambray and comparable fabrics are often dyed in this way.

■ **bales**

in countries where the fleece traditionally has been shipped, the fleeces are packed into bales, which load better in the ship holds for export abroad. Depending on the country, the bales weigh different amounts. Australian and New Zealand bales weigh 150 kg (330 lb), whereas South American bales weigh 1,000 lb (454 kg). Cotton also is shipped in 500-pound bales.

■ **baling press**

a machine used for reducing bolts of cloth or waste into compact bales for delivery.

■ **ball warping**

the winding up of enormous strands of yarns of a specified length and transforming him onto a beam in the form of a loose untwisted rope.

■ **ballotini**

small glass beads which are normally used in reflective paints but which can also be incorporated into fabrics.

■ **band**
commuting of spindle from cotton belt of textile machinery.

■ **bandana**
handkerchief designs in simple colour and white stylised patterns, including spots.

■ **bandanna**
a print design, done by effluence or resist printing, classified as white or brightly coloured motifs on a dark or bright ground, most often red or navy. It is generally done in India by tie-dyeing, on a cotton fabric.

■ **bandelier**
see **baldric**.

■ **bandhani**
a process of decorating a cloth by tie-dyeing in which on the undyed cloth by tying small spots very tightly with thread the design is reserved as the spot is unable to get the colour. It is especially popular in Rajasthan and Gujarat.

■ **banyan**
name given to an under-garment in Indian cloth. The term is refers to indoor garments 'dressing gowns'.

■ **barathea**
a twilled hopsack weave- with a fine texture, slightly pebbled surface. It is mostly of silk or silk blended with wool and can be used for neckties, women's fine suits and coats men's and women's evening wear.

■ **barathea**
a kind of rich soft-looking, fine fabric, which has granular texture achieved by the short broken ribs in the filling direction.

■ **bare pychon ka pyjama**
a pyjama with wide, loosened from the legs.

■ **bark cloth**
fabric made from the skin of trees or with a surface texture resembling tree bark.

■ **barras**
a coarse linen fabric similar to sackcloth, which is produced in Holland.

■ **barre**
an imperfection characterised by a ridge or mark running in the crosswise or lengthwise directions of the fabric. Barr's can be caused by tension variations in the knitting process, poor quality yarns and problems during the finishing process.

■ **barrier (geotextiles)**
a material that prevents fluid movement across the plane of a geotextile. A non-woven geotextile

saturated with an impermeable substance (e.g. bentonite clay) can act as a barrier material.

■ **barrier fabric**
fabric that proves to be obstructions to dust, dust mites and allied allergens.

■ **bas de cotte / de jupe / de robe**
a kind of term used in 17th century for the fabric used with lower part of the petticoat or skirt, which went with it, covered by the gown.

■ **basic dye**
1. an ion dye characterised by its substantivity for basic-dye able acrylic and basic-dye able polyester fibres, especially the former.
2. the first of the many groups of synthetic dyes. Sir William Perkins discovered them in 1856. Used on cotton, paper and wool. Provide brilliant shades but rated poor to fair to light and wash-fastness, colour resistance is only fair.

■ **basin waste, basineés**
the silk waste consisting of cocoons that could not be completely converted into bobbins because of frequent rifts in the thread.

■ **basket stitch**
a knit construction with mostly purl loops in the pattern courses to give a basket weave look.

■ **basket weaves**
a variation of the plain weave construction, formed by treating two or more warp yarns and/or two or more filling yarns as one unit in the weaving process. Yarns in a basket weave are laid into the woven construction flat and maintain a parallel relationship. Both balanced and unbalanced basket weave fabrics can be produced. Examples of basket weave construction include monk cloth and oxford cloth.

■ **basques**
the jackets with basques worn in combination with skirts instead of gowns. French word for short tabs at waist and male doublets that extended below the corset.

■ **bast fibre**
strong, soft, woody fibres, such as flax, jute, hemp and ramie, which are obtained from the inner bark in the stems of certain plants.

■ **batik**
process of dyeing in which portions of fabric are coated with wax and therefore resist the dye. It's a traditional Indonesian dyeing process and can be repeated to achieve multi-colour designs. Fabric generally gives a barred appearance wherever the dye enters through the cracks in the wax.

■ **batiste**
named for Jean Baptiste, a French linen weaver. The term is used as:

1. in cotton, it is a sheer, fine combed, mercerised muslin characterised by wide streaks. Woven of combed yarns, given a mercerised finish. Used for blouses, summer shirts, dresses, lingerie, infants' dresses, bonnets and handkerchiefs.
2. made of rayon and polyester and cotton blends.
3. made of wool or worsted in a smooth, fine fabric that is lighter than challis, very similar to fine nun's veiling. Used for dresses and negligees.
4. a sheer silk fabric, either plain or figured, very similar to silk mull. Often called Bastiste de Soie. Used for summer dresses.
5. made of spun rayon.

■ **batt (batting)**

sheets or rolls of carded cotton or wool or other fibre or mixtures thereof which is used for woollen spinning or for stuffing, padding, quilting and felting.

■ **batten**

see **beat up**.

■ **battery**

an arsenal on the loom that holds the full quills, cops, or bobbins of filling yarn and from there insertion into the shuttle takes place by an automatic changing device.

■ **bave**

a silk fibre withdrawn from a cocoon completely with its natural gum (sericin) as it is. It comprises of two brines.

■ **bayadère**

a fabric or design with horizontal plain or patterned stripes. Its generally bright coloured often in black warp and the effects are usually bizarre or startling. The name is of Indian origin and can be used in blouses or other dresses.

■ **BCF (Bulked Continuous Filament)**

textured yarn used mainly in the construction of carpets or upholstery.

■ **bead yarn**

a yarn upon which is fastened either an actual bead or (commercially) a lump of hardened gelatine of a bead-like form.

■ **beaded**

fabric ornated with beads.

■ **beading lace**

a lace manufactured with machines and a row of openwork holes intended for the insertion of a decorative ribbon.

■ **beam**

1. a large spool or roll, about three feet in diameter, on which warp or cloth is wound.
2. to wind yarn from a dyed ball warp onto a section beam.

■ **beam dyeing**

the warp is dyed prior to weaving. It is wound onto a perforated beam and the dye is forced through the perforations thereby saturating the yarn with colour.

beam warping

etching of yarn from bobbins or reels onto a warp or section beam in the form of a wide sheet. Various beams run through the slashing machine and then a loom beam is formed.

beaming machine

a machine, which helps in the process of winding the individual yarn, ends from a rope-like bundle and dispensing them evenly over a section beam.

beat up

to align and push the strands of filling yarn, closely together as they are woven. Reed is achieved by advancing and receding from the cloth, driving each pick against the fell of cloth already woven.

beaver

soft, silky, shiny, lended for textile use. Also, used in manufacturing of fur coats, trimming fur and fabric garments.

beaver cloth

an improved quality, heavy but soft in touch kind of soft wool cloth with a deep, smooth nap. It has a luxurious look and has the longest nap of all the napped fabrics and usually somewhat silky. It's often used in overcoats, caps maritime dresses and even shoes.

bedford cord

a cord cotton-like fabric with raised ridges in the lengthwise direction. Also made in wool or silk and rayon. Since the fabric has a high strength and a high durability, it has a very pronounced rib and firm construction. It is often used for upholstery, suitings, coatings and work clothes.

Cotton Shirt | Cotton Pants | Cotton Shirt

beet

a bale or sheaf of tied flax crop or straw.

beetle

a large wooden mallet used to help soften cellulose fibres. Often used with linen and ramie.

beetled

accomplishing process in which a fabric is pounded with sheen to produce a hard flat surface.

beetling

the process of striking woven linen or ramie fabric with rollers to flatten the fibres. This leaves you with a more lustrous fabric.

belly wool

the wool that grows on the belly of the sheep and occasionally extends up the side in irregular patches. It is usually an uneven, different grade from the body of the fleece. It is shorter and less desirable because of its poor lock formation and it usually lacks the character of the body of the fleece.

■ **belt-edge separation (tyres)**

separation of the plies of reinforcing fabric from the rubber matrix of a tyre, at the edge of the belt of reinforcement.

■ **bengaline**

a sturdy, warp-faced fabric with pronounced crosswise ribs formed by bulky, coarse, plied yarn or rubber threads. Filling is not discernible on face or back of goods. Originating in Bengal, India, it is used in coating, swimsuits, mourning material, ensembles and women's headgear. Grosgrain is 'bengaline cut to ribbon width'. It is used in coats, suits, millinery, trims, bouffant dresses with a tailored look, mourning cloth and draperies.

■ **bias**

flow of a fabric in any direction and not exactly in the direction of the vertical yarns or horizontal yarns of a fabric. An accurate bias makes an angle of 45 degree across the length and width of a fabric, fabric cut on a bias has maximum stretch.

■ **bias belted tyres**

tyres reinforced by layers of tyre cord fabric arranged alternately so that the main load bearing yarns lie at an angle of less than 90° to the plane in which the tyre rotates and yarns of adjacent layers cross each other.

■ **bicomponent fibre**

a man-made fibre with two distinct polymer fibre forming components. Certain fibres are considered as bicomponent like wool and some other animal fibres because they obsess.
A side-by-side configuration of the ortho- and para-cortex resulting in crimp in the fibre.

■ **bicomponent yarn**

a yarn having two different continuous filament components.

■ **bilaminate (fabric)**

a fabric formed by bonding two separate fabrics together.

■ **binders**

the individual hairs in a sheep's fleece that run from one staple to another.

■ **binding threads**

threads used to unite two or more ply into one firm (stable) structure.

■ **biocompatibility**

compatibility with living tissue or a living system by not being toxic or injurious.

■ **birds eye**

1. a general term for a fabric with a surface texture of small, uni-

form spots that suggest bird's eyes. Can be woven or knit.
2. a design that suggests a bird's eyes.

■ **biscuit**

one of the several narrow cylindrical cheeses of yarn wound as a composite package on a single former side by side but not touching.

■ **bi-shrinkage yarn**

a yarn containing two different types of filament, which have different shrinkages.

■ **black wool**

any wool containing black fibres. A fleece having only a few black fibres is rejected by a grader and goes into the black wool bag because there is no way of separating the few black fibres in the manufacturing processes. Black wool is usually run in lots that are to be dyed.

■ **black-top wool**

wool containing a large amount of wool grease combined at the tip of the wool staples with dirt, usually from a Merino. This wool is usually fine in quality, of good character and desirable in type, but the shrinkage is high.

■ **blanket cloth**

a soft, raised finish, 'nap' obtained through gliding by the fabric over a series of rollers covered with fine wire or teasels. Heavily napped and depressed on both sides. The matter is used in bed covering, overcoats and robes.

■ **blanket plaid**

a grand vividly coloured chequered design such as those often found on blankets.

■ **blaze**

see **cocoon stripping**.

■ **bleaching**

the procedure of enhancing the whiteness of textile material, with or without the effect on the natural colouring matter and/or extraneous substances, by a bleaching agent.

■ **bleaching agent**

a chemical reagent capable of making the texture white by destroying partly or completely the natural colouring matter of textile fibres, yarns and fabrics. Hydrogen peroxide is the best example and among others oxidising and reducing agents.

■ **bleeding**

a term applied to yarn from which the colour runs, usually staining the white or lighter coloured-items nearby.

■ **blend**

a term applied to a yarn or a fabric that is made up of more than one fibre. In blended yarns, two or more different types of staple fibres are twisted or spun together to form the yarn. Examples of a typical blended yarn or fabric is polyester/cotton.

■ **blending**

a procedure concerned primarily with efficient mixing of various lots of fibres. It is normally carried out to mix fibres, which are of different physical properties, market values or colours.

■ **blending machine**

a group of devices that are used in the process of blending. The machine is synchronized to definite amounts of various grades of cotton, which are to be blended together.

■ **blinding (geotextiles)**

a condition in which soil particles block openings on the surface of a geotextile, thereby reducing the hydraulic conductivity of the geotextile.

■ **blitz**

it's a woven fabric with a filament warp and spun weft that is light to medium in weight. It often has a very fine crosswise rib. Some common blends are acetate/rayon and polyester/rayon.

■ **block copolymer**

a copolymer in which the repetition of units occur in blocks.

■ **block printed**

a hand printing method with the help of wood, metal, or linoleum blocks. One block for one impression and colour. The dye is applied to the block, which is pressed or hammered against the fabric.

■ **block printing**

the oldest form of printing known to man. Motifs are obtained by the use of wooden, linoleum or copper blocks. This operation is very tedious, production is very low,

prices are rather high and there has to be a separate block used for each colour chosen by the designer.

■ **blocker**

a frame for drying wool. This is an open frame that rests on two supports with a handle on one side. You wind the damp yarn under even tension across the frame— not trying to line anything up. Rather like winding a bobbin for weaving. After the yarn dries, you can usually slide the whole skein off of one end.

■ **blocking**

the process of drying a skein of wool under tension. Drying a skein on a blocker can do this. More prosaically, it can be done by winding around an upended-chair's legs or by hanging a weight in the bottom of a skein.

■ **blood (blood grade)**

this refers to the fineness of the wool, measured as low 1/4, 1/4, 3/8 and 1/2 blood. It reflects the amount of Merino blood in a breed. 'More blood' refers to a larger amount of Merino in a sheep, which should produce a finer wool.

blotch print

refers to the printing uniform colour in a larger area and the printed ground is referred to as the blotch.

blowing room

the room where the preceding processes of budding, cleaning and blending are carried out in a cotton spinning mill.

blowout factor

the rapidity with which an animal's fibre diameter thickens with age.

bobbin

the cylinder or spool upon which yarn or thread is wound. An option to buying lots of bobbins for your wheel is using 'Storage Bobbins'.

bobbin cleaner

a machine that helps in removal of any remaining yarn or roving from bobbins (quills) after they have been used in the looms, spinning machines or winders.

bobbin lead

a single band drives the bobbin. The flyer has a friction brake. A well-known example of this would be the Ashford Traditional. Another term used is 'Scotch Tension'.

boculè

a compound yarn comprising a twisted core with an effect-yarn wrapped around so as to produce loops on the surface.

bod

biological oxygen demand-a measure of pollution by oxygen-consuming organic materials in an effluent stream.

body

a term applied to wool when the staple has a good 'hand' (full and with bounce). It can also refer to the fullness of a fabric. This is a subjective quality and has to do with a lack of limpness and/or stiffness. A fabric is said to have a good body when it has a full, rich and supple hand.

boiled wool

thick, dense fabric that is heavy to completely obscure its knitted construction.

boiling

a process in which a yarn or garment made from staple fibre containing wool or animal hair is left in boiling water so that the original fabric construction is obscured by the felted surface.

boiling off

the operation of removing, by means of a hot, mildly alkaline liquid, the gum (seracin) that covers the raw silk fibre. Also called 'de-gumming'.

bolivia (elysian)

a pile weave (cut) with a diagonal pattern. It has a soft and velvety feel because it sometimes contains alpaca or mohair. Usually piece dyed it has lines or ridges in the warp or in a diagonal direction on one side and comes in light, medium and heavy weights. It is used in cloaking, coatings and some suits.

■ **boll**
a seed case and its contents, as of cotton or flax.

■ **bolt**
see **pieces**.

■ **bolt of cloth**
a rolled or folded length of cloth.

■ **bombazine**
a kind of very fine English fabric, usually has silk or rayon warp and worsted filling. Title is derived from a Latin word 'bombycinum' that means a silk in texture. It is one of the oldest materials known and was originally all-silk. It is generally used in Infants wears. When dyed black it is used in the mourning cloth trade.

■ **bonded fabric**
a non-woven fabric in which a bonding material holds the fibres together. This may be an adhesive or a bonding fibre with a low melting point. Alternatively, stitching may hold the material together.

■ **bonding agent**
see **binders**.

■ **book**
a parcel of hunks of raw silk weighing around 2 kg.

■ **border**
a design placed along the extremity of the fabric or at the edge

designed in such a fashion that it will fall on the edge of the finished product and will enhance its appearance. These designs are frequently used in skirts, sarees and dresses.

■ **botanical**
referring to designs dominated by motifs depicting plant life.

■ **botany wools**
originally referred to merino wool shipped from Australia's Botany Bay. It has become a generic term used to describe superlative wools and fine worsted sweaters.

■ **boucle**
1. a fancy yarn with an irregular pattern of curls and loops.
2. a fabric made from boucle yarn.

■ **bouclette**
a small bouclé effect.

■ **bourdalou**
hat-ribbon, finer than grosgrain, rounded around the foot of the crown of hats. This trimming is in use since the 17th century.

■ **bourdon lace**
a machine made lace on a mesh ground usually in a scroll design outlined with a heavy cord.

■ **bourette**
a silk noil fabric made from short fibre (silk waste) with a textured surface.

■ **bourrelet**
a double knit fabric with a rippled corded texture running horizontally.

bowl
one of a pair of large rollers forming a nip.

boxtruck
a box used for transmitting articles like bobbins and spools from one department of a mill to another.

bradford count (bradford system)
the British standard is based on the Bradford Spinning Count System. This originated in the 19th century and is based on the number of 560-yard worsted skeins that can be produced from one pound of clean wool. The clean wool is then thoroughly oiled which aids in producing a smooth, lustrous yarn for suiting. With this system the larger number will be finer wool.

braid wool
the coarsest of the U.S. grades of wool, according to the blood system of classification. It is very coarse and lustrous wool.

braided yarn
intertwined yarn containing two or more strands.

brandenburg coat
a loose overcoat with turned-back cuffs and sleeves made in one with the rest of the garment.

break
weak at a certain point, but strong above and below the weak spot, as opposed to 'tender', which signifies a generally weak fibre. This can be caused by a sudden change in pasture, feed, illness or lambing.

breaker picker
the first between the two units of older style picker machines. In this unit the cleaning of raw cotton is partially done by beating and fluffing and then fed into a finisher picker.

breaking
also known as 'scratching'. In breaking, the flax plants that have been through the 'retting' process pass through rollers or are beaten with a wooden blade to help 'break' the stronger parts without damaging the longer fibres.

breaking and opening machine
a set of machines that tears apart and partially cleans matted, compressed and baled cotton.

breaking extension
the percentage extension at maximum load.

breaking length
a measure of the breaking strength of a yarn. It is the calculated length of yarn, which equals its breaking load and is equal to the tensile stress at rupture of the yarn.

breaking load, breaking force
the load that develops the breaking tension. The unit of measurement sanctioned is the Newton.

breaking strength (geotextiles)
the ultimate tensile strength of a geotextile per unit width.

■ **breaking stress**

the maximum effort developed in a specimen stretched to rupture. If the actual stress, defined in terms of the area of the strained specimen, is used, then its maximum value is called the actual breaking stress.

■ **breaking tension**

the maximum force developed to rupture in a specimen. It is correctly expressed in Newtons.

■ **breast beam**

the bar, at front of the loom, that guides the woven cloth onto the cloth roll.

■ **breathability**

the ability of a fabric, coating or laminate to transfer water vapour from one of its surfaces through the material to the other surface.

■ **breathable coated**

it refers to a kind of coating that not only repels water but also allows water vapour (thus perspiration) to pass through, allowing garments to be comfortable and waterproof. It is used in garments for active wear and winter sports.

■ **breech (britch wool)**

wool from the thigh and rear region of the sheep. It is the coarsest and poorest wool on the entire fleece. It is usually manure-encrusted and urine-stained fibre. It should be 'skirted' and removed from a fleece for a hand spinner.

■ **breed characteristics (breed type)**

individual breeds have distinct characteristics. A Merino is very fine, shows a lot of crimp and the fibres are very close. A Lincoln, is much coarser with low crimp.

■ **breton lace**

lace embroidered on an open net with heavy often brightly coloured yarn. It can be hand made or machine made.

■ **bright**

very white, almost reflective, wool relatively free of dirt and sand. Some breeds, like Cormo, are known for producing particularly bright fleeces.

■ **britch**

this is the short, curly fibres found in the groin and belly area of sheep. It has a very different character from the rest of the fleece and should be skirted out. In a perfect world, spinners would never see this.

■ **brittle**

brittle refers to harsh, dry, 'wire-like' fibre, much like the split ends in hair.

broadcloth

1. originally a silk shirting fabric, so named because it was woven in widths exceeding the usual 29 inches.
2. a tightly and plain-woven lustrous cotton cloth with a crosswise rib, resembling poplin, but ribs are finer and broadcloth has more picks than poplin. Finest qualities are made of combed Pima or Egyptian cotton. Used for men's shirts, women's tailored dresses and blouses.
3. smooth rich-looking woollen with napped face and twill back. Better grades have a glossy and velvety feel. Used for dresses and skirts.
4. also made of a blend of cotton and polyester and/or other manmade fibres.

brocade

rich Jacquard-woven fabric with all over interwoven design of raised figures or flowers. Name derived from the French word meaning to 'ornament'. Pattern is emphasised by contrasting surfaces or colours. Often has gold or silver threads running through it. It's rich, heavy and elaborates design effects. Background may be either satin or twill weave. The motifs may be of flowers, foliage, scrollwork, pastoral scenes, or other designs. Used for dresses, wraps, draperies and upholstery, depending on the weight.

Marlene's
ROSE BROCADE

brocatelle

a heavy figured cloth in which the pattern is created by warp threads in a satin weave. It is recognised by a smooth raised figure of warp-effect, usually in a satin weave construction, on a filling effect background. True brocatelle is a double weave made of silk and linen warp and a silk and linen filling. It is generally used in decoratives, draperies and furniture coverings.

broken end

a thread or strand of cotton, which has broken in a textile machine.

broken twill

a general term for twill weave fabrics in which the twill line changes direction.

brushed wool

finished yarn or material that has been brushed to raise all loose fibres to the surface, i.e., the commercially spun mohair yarns.

brushstroke

refers to a print style in which colour looks are applied with a brush.

brussels lace

may be a bobbin or needlepoint lace usually on a machine made ground. Sometimes designs are appliquéd on the ground. As Brussels Belgium is important in the history of lace making, many different types of lace are called Brussels lace.

buck fleece

a fleece from a ram. The wool usually has heave shrinkage due to excessive wool grease, thus

Textile   buckram | burn out

wool of this type is not worth as much in the grease as similar wool from ewes or withers. Some buck fleeces have a distinctive odour that many find objectionable.

■ **buckram**

1. ply yarn scrim fabric with a stiff finish.
2. made by gluing two open weave, sized cotton fabrics together. Used as interlining in cloth and leather garments. Also in millinery because it can be moistened and shaped.

■ **buckskin**

a heavy satin weave fabric with a glazed face, generally of fine merino wool.

■ **buffalo check**

a bold check pattern with blocks of 2 or 3 contrasting colours. Colours used are often red and black in a twill weave.

■ **bulk grade**

the largest percentage of grade in a lot of original-bagged wool.

■ **bulked yarn**

a yarn that has been treated mechanically, physically or chemically so as to have a noticeably greater voluminosity or bulk.

■ **bulky**

in wheels, a term used for a wheel with a wide orifice. This allows the creation of a thicker yarn suitable for blanket wefts.

■ **bump**

a cylinder of coiled, prepared fibres ready for spinning. This is

how commercially prepared fibres are delivered. Rather if you had access to a really big ball winder and used it to wind the top you had just hand combed.

■ **bunch (flax)**

the tied up aggregate of pieces, with two or more ties preparatory to baling.

■ **bunting**

the name derived from the German bunt, meaning bright. Cotton or worsted yarn is used to make this soft, flimsy, plain-woven cloth. Some of the cloth is made from cotton warp and worsted filling. Cotton bunting is made from heavy cheesecloth and comes in the white or is piece-dyed. Ply yarn may be used in this plain-woven, fairly loose-textured cloth, which has a texture ranging from 24 x 36 to 24 x 32. All-worsted bunting is used in making flags.

■ **burl (speck dyeing)**

done mostly on woollen and worsteds, coloured specks and blemishes are covered by the use of special coloured inks, which come in many colours and shades. This is a hand operation.

■ **burlap/hessian**

a coarse open fabric constructed of jute that is often used for upholstery lining and bagging. When dyed or printed it is used in drapery, wall coverings, upholstery.

■ **burn out**

a brocade-like pattern effect created on the fabric through the application of a chemical, instead of colour, during the burnout printing process. (sulphuric acid, mixed into a colourless print paste, is the most common chemical used.) Many simulated eyelet effects can be created using this method. In these instances, the chemical destroys the fibre and creates a hole in the fabric in a specific design, where the chemical comes in contact with the fabric. The fabric is then over-printed with a simulated embroidery stitch to create the eyelet effect. However, burnout effects can also be created on velvets made of blended fibres, in which the ground fabric is of one fibre like polyester and the pile may be of a cellulose fibre like rayon or acetate. In this case, when the chemical is printed in a certain pattern, it destroys the pile in those areas where the chemical comes in contact with the fabric, but leaves the ground fabric unharmed.

■ **burry wool**

wool heavy in vegetable matter including burs, leaves, seeds and twigs, which requires special and expensive processing in removal.

■ **bursting strength**

the mechanical test done commercially on fibres to show how strong they are.

■ **buta**

a floral motif, derived generally from Persian sources, much used in Indian textile design and traditionally rendered as a flowering plant with a curling bud at the top. The motif is also sometimes reduced to a floral pattern designed within the form of the plant.

■ **butcher's linen**

it was originally made with linen but is now created with cotton or manufactured fibres. A strong, heavy, plain weave linen fabric with uneven, thick and thin yarns in both warp and weft - often used in tablecloths and aprons.

■ **buti**

a trivial of buta, very commonly used in Indian textile design.

■ **butt**

to straighten the root ends of flax straw at different stages of processing by quivering it upright on a flat surface, either by hand or mechanically.

■ **butter muslin**

see **muslin**.

■ **cabin**

an organised compartment which works as a storehouse for the filling yarn of various sizes, colours and quality.

■ **cable**

to twist together two or more folded yarns.

■ **cable stitch**

a knit fabric stitch that produces a design that looks like a heavy cord- common in sweaters and hosiery.

■ **cabled yarn**

two or more plied yarns twisted together, one or more part of a cabled yarn can be a single. So if you took two 2-ply and plied them again, you would have a cabled yarn. It is important to remember that you reverse the twist for each step. So if you spun your singles z, the 2-ply would be spun s and the cabled yarn would be produced by plying z. You will need to have extra twist in the singles and the first ply to produce a 'balanced yarn'. A 3-2 cable refers to three 2-ply.

CAI

Compression Strength After Impact.

cake

the cylindrical package, of continuous-filament yarn produced in the viscose spinning industry by means of a top ham box.

calendar rolls

a unit on the sliver lapper, ribbon lapper and combing machine which presses the ribbon lap or sliver, as it comes from the drawing rollers, into a loosely matted layer.

calendared

the term used to describe a fabric which has been passed through rollers to smooth and flatten it or confer surface glaze.

calendaring

a process for finishing fabrics through a calendar in which such special effects as high lustre, glazing, embossing and moiré are produced. It helps in producing a glaze on the face of the fabric that is in contact with the steel bowl.

calico

originated in Calcutta, India, it is one of the oldest cotton staples on the market. This plain and closely woven, inexpensive cloth is made in solid colours on a white or contrasting background. Very often, one, two, or three colours are seen on the face of the goods, which are discharge or resist-printed. Calico is not always fast in colour. Medium yarn is used in the cloth and the designs are often geometric in shape. The yarn used is 30s and the texture is about 66 x 54. Uses are for aprons, dresses and crazy quilts. Interchangeable with percale.

cambric

a plain weave, traditionally lightweight cotton fabric with lustre on the surface. Used for handkerchiefs underwear, shirts, aprons, tablecloths. Highly mercerised and lint free, calendared on the right side with a slight gloss. The lower qualities have a smooth bright finish. It is used in night gowns, handkerchiefs, underwear, slip, aprons, shirts and blouses.

camel

the hair of the camel or dromedary, also used as a broad description of fawn colour.

camel hair

the hair of the camel (camelus bactrianus) or dromedary. It consists of the strong, coarse, outer hair and the undercoat. The better grades the more they are expensive. Sometimes blended with wool to reduce the cost and increase the wear. It is used for over-coating and top coating.

camelid

any animal that comes from the camel family. obviously camels, but also alpaca and llama.

camocas

a very beautiful fabric that was often stripped with gold or silver with a satin base and was popular in the 14th and 15th centuries. It is diapered like fine linen.

can

a large cylindrical container for receiving and holding lengths of sliver delivered from the front of a carding machine, drawing frame, or combing machine.

canary-stained wool

a yellowish colouration in the wool, which cannot be removed by ordinary scouring methods. May be caused by bacterial growth or urine staining.

candle

this refers to the stiffened fat on an unwashed fleece. Not a pleasing condition for hand spinners and often a condition when fleeces sit for years waiting to be spun.

candlewick

a mossy pile fabric with a fuzzy surface that looks like chenille. Bending a heavy plied yarn on a muslin base then cutting the loops makes it. It's frequently used for bedspreads, robes, draperies.

canons, also cannons

17th century, full, wide ruffles/flounces attached at the bottom of breeches, especially petticoat breeches. It was a sort of half-stocking, at first long and narrow, then wider and decorated with flounces and lace.

canton flannel

a heavy, warm, strong cotton or cotton blend fabric with a twill face and a brushed back. Used for nightwear, underwear, gloves, linings. It is originally produced in canton China.

canvas

1. cotton or linen fabrics with an even weave that is heavy and firm.
2. ada or java canvas is a stiff open-weave fabric used for yarn needlework.
3. awning stripe canvas has painted or woven stripes on cotton duck.
4. cross-stitch or penelope canvas is used for fine cross-stitch patterns, has stiff, open mesh.

cape net

a stiff heavy net that can be moulded in any shape when wet and it sustains that shape. It is used in hats.

capotain or copotain
a grand conical, crowned and small-bounded cap fashionable in the 16th century-17th century.

caprolactam
a chemical intermediate used in the manufacture of polyamide (nylon).

carbon (fibre) (generic name)
a term used to describe fibres that contains at least 98% of carbon obtained from controlled pyrolosis of appropriate fibres.

carbonising
the process of treating wool with chemicals, usually acids, to destroy and remove the burrs without seriously damaging the wool. The usual chemical used is sulphuric acid. Wool so treated is known as carbonised wool.

carbonised rag fibre
animal fibre derived from either the wet or the dry carbonising process.

card
a tiny unit of a archetype chain used on a jacquard loom. It is a cardboard strip with holes (resembling those in a player piano roll that are punched) that serves to control the action of the weaving mechanism.

card clothing
It is a kind of special cloth or rubber, studded with wire teeth, which helps to open up the cotton fibers and clean the impurities and align them in a parallel order. The material is affixed to the various working parts of the carding machines.

card cutting
the act of punching holes in jacquard cards as per design craft, with the purpose of controlling weaving mechanism and the pattern will be woven into the cloth.

card cylinder
1. that part of a jacquard loom, which holds the pattern, cards in position while it gambles and also helps in controlling of the weaving of patterns when they pass through the holes in cards.
2. see **carding drum**.

card grinding
timely sharpening of the wire bristles of the carding machine with an abrasive cylinder.

card punching machine
a machine that helps in checking the weaving of designs and patterns in the cloth by penetrating in the cards that are used in jacquard looms.

card sliver
a thick, disentangled rope of cotton fibres, which represents the finished product of the carding machine. It is uniform in thickness and relatively free from naps.

carded
description of a continuous web or sliver produced by carding. A yarn in which the fibres are partially straightened and cleaned prior to spinning. The yarn is generally coarser and more uneven than a combed yarn.

carded fibres

fibres that have been carded which opens them up.

carders

also known as hand carders (as opposed to 'drum carders'). Some of the carders have curved backs, some straight backs. There is some belief that the reason why modern hand cards have the curved backs is because they were modelled after museum pieces. Unfortunately, the museum pieces were warped (curved). Early plans for carders show the straight backs.

cardigan -full

a kind of fabric which has variation of a 1x1 rib stitch with 2 sets of needles and the process takes place with alternate knitting and tucking on one course then tucking and knitting on the next course. Also famous as polka rib, the fabric contains the similar look from both the sides and has both a held loop and a tuck loop.

cardigan- half

fabric with variation of a 1x1 rib stitch with knitting and tucking in alternate courses on one set of needles. Its also known as royal rib and the construction on the back is the reverse of the face.

carding

carding is the process used to open out fleece so that it can be more evenly spun into a 'woollen' yarn. The process by which the fibres are opened out into an even film. It is the process of separating and cleaning cotton fibres to prepare them for spinning.

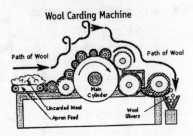

carding cloth

the woolly designs site has a close-up of carding cloth. The material is used on hand cards, drum carders and carding boards. The spacing of the tines causes it to be classified as 'fine' or 'coarse'. Many manufacturers refer to their combs as 'cotton cards' or 'wool cards'.

carding drum

a large rapidly revolving cylinder of the carding machine, which is covered with several million wire teeth, that pulls out the cotton fibers and crushes them to remove small particles of dirt and knotted fibers.

carding machine

a machine that transforms the cotton from lap into sliver by first completing the cleaning of the cotton and arranging the fibres so that they are mostly parallel.

■ **carding wools**
wools that are too short to be treated by wool combing and must be processed into woollen yarns. synonymous with 'clothing wool'.

■ **carpet beetle**
the larva of this beetle eats wool and other protein fibres.

■ **carpet wool**
coarse, harsh, strong wool that is more suitable for carpets than for fabrics. Very little of this type is produced in the U.S. some of the choicer carpet wools are used to make tweeds or other rough sport clothing. Some breeds, like karakul, are mainly used for rugs.

■ **carrier (colouration)**
a kind of accelerant, especially used in the dyeing and printing of hydrophobic fibres with radiated dyes.

■ **carrotting**
the modification made in the tips of fur fibre (rabbit fur) by chemical treatment with the purpose of improving their felting capacity. Mercury in nitric acid and mixtures of oxidising and hydrolysing agents are certain important reagents used in the process.

■ **casein**
the main protein in milk that also serves as raw material for some mended.

■ **casement clothe**
a term used for sheer, lightweight, open weave fabrics used in the making of curtains and backing for heavy drapery.

■ **cashgora**
fibre produced by crossing cashmere goats with angora goats.

■ **cashmere**
soft, silky fibre combed from the cashmere goat with a diameter of 18.5 microns or less. Cashmere has excellent insulating power, providing warmth without weight or bulk. It drapes beautifully, resists wrinkles and sheds lint. Costly because of limited supply. Its derived from the Kashmir goat, a hair fibre found in India, Tibet, Iran, Iraq, China, Persia, Turkestan and outer Mongolia. Often mixed with wool or synthetics to cut costs and improve the wear. The textile industry is only interested in the soft fibres. Knitted into sweaters for men and women, also women's dresses. Often combed and sold in tops and noils.

■ **cassock casaque**
a full-sleeved three-quarter length coat cut with wide, coat running throughout the body, ending at thigh-height or below. It is an unbelted overcoat, open-sided and almost always covered with braid and woven ornament.

castle wheel

the flyer is usually mounted above the wheel, which means less floor space is used. A well-known example of this is the reeve's castle wheel.

cationic

a type of dye, which is generally used on acrylic or on, modified polyester or modified nylon yarn. It is often used to achieve cross-dyed effects by weaving dye able yarn in a pattern with regular yarn in the same fabric. The pattern is visible by dyeing the fabric in 2 baths.

causticising

a brief treatment of cellulose fabrics with caustic soda solution at room temperature, particularly with reactive dyes.

cavalry twill

a firm warp-faced cloth, woven to produce a steep twill effect. A kind of sturdy woven fabric with a steep pronounced double twill line. It is often of cotton or wool but may be any fibre.

cellophane effect

an effect created in a fabric that gives it the iridescent appearance of cellophane.

cellulose

a material derived from the cell walls of certain plants. Cellulose is used in the production of many vegetable fibres, as well as being the major raw material component used in the production of the manufactured fibres of acetate, rayon and triacetate.

cellulose fibre

fibres produced from the cell walls of plants, i.e., cotton, hemp, ramie.

cellulose triacetate

in theory it deals with cellulose acetate which contains 62.5% of combined ethanoic acid (acetic acid) but the term is also considered as used for primary cellulose ethanoate (acetate) containing more than 60% of combined ethanoic acid.

cellulose xanthate

a sequence of compounds that are formed between carbon disulphide and cellulose in the presence of strong alkali.

cellulosic filament

filaments made or chemically derived from a naturally occurring cellulose raw material.

cendal

made in various qualities it is a kind of silk material resembling taffeta. Also famous as a luxury fabric, sometimes only as cheap lining material.

centinewton

a unit of force used to measure the strength of a textile yarn.

centipoise

a measure of viscosity, equal to 0.001 Newton second per m^2.

centre front

that portion of the pattern or the garment, which is comes exactly in the front.

Textile — centrifugal spinning | chameleon — 35

■ **centrifugal spinning**

process by which a man-made fibre production in which the molten or dissolved polymer is thrown centrifugally in fibre form the edge of a surface rotating at high speed. The term is also referred to describe a method of yarn formation by rotating of cylindrical container, in which, the yarn passes down a central guide tube and is then carried by centrifugal force to the inside of a rotating cylindrical container.

■ **CFRP**

Carbon fibre reinforced plastic.

■ **chaconne**

a kind of cravat manufactured from a ribbon dangling from the shirt collar to the chest. It is famous after the name of a dancer pécourt of 1692.

■ **chafer fabric**

a fabric coated with vulcanised rubber, which is wrapped around the bead section of a tyre before vulcanisation of the complete tyre. its purpose is to maintain an abrasion-resistant layer of rubber in contact with the wheel on which the tyre is mounted.

■ **chaff**

a heterogeneous assortment of vegetable fragments found as the component of trash in cotton, most of them being small pieces of leaf and stalk.

■ **chaffy wool**

wool containing a considerable amount of chaff: finely chopped straw.

■ **chain dyeing**

used when yarns and cloths are low in tensile strength. Several cuts or pieces of cloth are tacked end-to-end and run through in a continuous chain in the dye liquor.

■ **chainette**

a tubular cord produced on a circular knitting machine.

■ **challis**

a lightweight, soft plain weave fabric with a slightly brushed surface. It is lightweight, plain weave fabric with good drape. Often used for printed dresses and skirts. The fabric is often printed, usually in a floral pattern. Challis is most often seen in fabrics made of cotton, wool or rayon.

■ **chambray**

1. a plain woven fabric almost square count (i.e. 80 x 76), with coloured warp and white filling that gives a mottled coloured surface. used for shirts, children's clothes and dresses. named for Cambria, France, where it was first made for sunbonnets.
2. a similar but heavier carded yarn fabric used for work and children's play clothes.
3. usually plain but may be in stripes, checks or other patterns. often used in shirts, dresses children's clothes.

■ **chameleon**

a 3-tone effect that is achieved by using a warp yarn of one colour and double weft yarns of 2 different colours that changes with the angle of view. It is generally found in taffetas, poplins or faille's of silk or made filament yarns.

■ **chamois cloth**

a kind of cotton, plain fabric that is napped, sheared and dyed to simulate chamois leather. More stiffer than kasha and thicker, softer and more durable than flannelette. Also designated as 'cotton chamoise-colour cloth'.
Used in dusters, interlining, storage bags for articles to prevent scratching.

■ **chamoisette**

a fine, firmly knit fabric, basically cotton or rayon, which has a very short soft nap. Nylon chamoisette is more often called 'glove silk', as it is used in gloves.

■ **chand-tara**

its literally used in India, referring to 'moon and star', a pattern often used in textile.

■ **chantilly lace**

a spool lace carved on a fine net ground represented by delicate motifs of scrolls, vines, branches and flowers outlined by a flat (cordon net) yarn. It is generally black in colour, a make of chantilly France.

■ **character**

the evenness, distinctiveness and uniformity of crimp characteristic of their respective wool classes. A well-bred wool of 'good character' will usually show a pronounced crimp and distinct staple formation.

■ **charged system**

process of dry cleaning in which an oil-soluble reagent such as petroleum sulphonate is added to the solvent so that a significant amount of water can be added to obtain a substantially clear dispersion of water in the solvent. The concentration differs in high-charged system, reagent is added, of so-called detergent that is 4% while, in a low-charged system the concentration ranges from ¾% to 2%.

■ **charka**

charka (means wheel) was developed in India by Gandhi in early 1920's so the people of India could spin cotton thread and not is dependent on foreign materials. The book-size charka is a mobile, self-contained charka. Charkas are designed for spinning fine fibres such as cotton, silk, angora and cashmere, etc.

■ **charmeuse**

charmeuse is a satin weave silk with a crepe back sometimes called crepe-backed satin. It is a soft lightweight woven satin fabric with good drape. It is made with high twist yarns, has a semi-lustrous face and a dull back. Often used for blouses, intimate apparel.

■ **charvet**

it is a soft, silky fibre with high lustre and a warp face, woven in herringbone. Earlier it originated as a silk fibre but is now made of manufactured fibres.

■ **chaubandi chola**

a kind of short tunic or shirt fastened with tie-cords generally worn by children.

chaugoshia (topi)
a four-cornered cap.

chauri
a flywhisk made generally from a yak's tail, important as a symbol of royalty or divinity.

check
a small pattern of different coloured squares or rectangles. It may be printed, yarn dyed, cross-dyed or woven into the fabric.

cheese
a roll of yarn resembling bulk of cheese, built up on a paper or wooden tube.

cheese cloth
see **muslin**

cheesecloth
an open lightweight plain-weave fabric usually made from carded cotton yarns.

cheeses
cheeses refer to the spirals of pencil roving produced on the large mechanised carders. The fibres can be knit as is (the original lopi) or can be spun up.

chelate
a chemical compound whose molecules contain a closed ring of atoms of which one is a metal atom.

chelating agent
a chemical compound which coordinates with a metal to form a chelate and which is often used to trap or remove heavy metal ions.

chemic, chemick
calcium or sodium hypochlorite.

chemical bonding
part of a production route for making non-woven, binders are applied to a web which, when dried, bond the individual fibres to form a coherent sheet.

chemicking
blanching done by means of a dilute hypochlorite solution. It is often done on non-protein fibre material.

chemise
a light undergarment made from linen for both sexes.

chenille
a yarn consisting of a cut pile that may be one or more of a variety of fibres helically positioned around axial threads that secure it. Gives a thick, soft tuft silk or worsted velvet cord or yarn typically used in embroidery and for trimmings along with millinery, rugs, decorative fabrics, trimmings, upholstery.

cheviot
1. a rough, harsh surfaced fabric of wool with a heavy nap. Used for coating.

2. a loosely woven tweed fabric with a shaggy texture originally manufactured from the wool of the cheviot sheep.
3. used in making of coats, suits, sportswear, sport's coats.

■ **chevron**

a design, which incorporates, herringbone elements of zigzag stripes or joined Vs.

■ **chiffon**

this term implies thinness, diaphanous or gauze-like and softness, as well as strength in a flimsy fabric. Its strength is brought about by the use of s twist in warp and filling yarns. Made of 50-denier silk yarn, it is finished around 47-inches and has textures of 60 x 60 up to 80 x 80. Georgette crepe is made with two ends and two picks of s-twist and z-twist while chiffon, the lighter fabric is made the one twist in it. Uses include dress goods, eveningwear, lampshades, millinery, trimming and underwear.

■ **chikan kari**

one of the famous works of Lucknow embroidery done on white cotton thread upon fine white cotton fabric, like, muslin.

■ **children's**

refers to designs suitable for the children's market.

■ **child's pudding**

a kind of protection done in form of small round hats for children made of cloth or straw.

■ **China grass**

an alternative name for ramie, a bast fibre.

■ **China silk**

originally hand woven in China of silk from the bonabyx mori. It is very soft and extremely lightweight but fairly strong type of silk. It is mostly used for linings and under linings and could be used for blouses.

■ **chinchilla**

a thick, heavy, pile fabric with surface curls or nubs, originally made to suggest chinchilla fur. It is often double faced and may be woven or knit and is often used as coating. It attacks the face and causes the long floats to be worked into nubs and balls. Mostly used in cotton, used for baby's blankets and bunting bags.

■ **chine**

textiles with a mottled pattern.

■ **chino**

a sturdy, medium weight, twill fabric usually of cotton or a cotton blend. Fabric is mercerised and sanforised, washes and wears extremely well with a minimum of care. It has often been used for summer weight military uniforms, sportswear and work clothes. It is often found in khaki and tan colours. Used in army uniforms, summer suits and dresses, sportswear.

■ **chinoiserie**

fabric designs which are derived from or which are imitations of Chinese motifs.

■ **chintz**

glazed cotton fabric often printed with gay figures and large flower designs. Named from Hindu word meaning spotted. Some glazes will wash out in laundering. The only durable glaze is a resin finish that will withstand washing or dry cleaning. A plain weave fabric, usually cotton, with a multicolour print which may or may not be glazed. If it is unglazed it is called cretonne. Used for draperies, slipcovers, summer dresses and skirts.

■ **chirimen**

a dull crepe fabric made with a course yarn, it is a Japanese term describing. Originally of silk but now found in man-made filaments such as polyester.

■ **chite**

painted linen originally from Chital (India), where the trend of painted linens was started in the 17th and 18th centuries.

■ **chlorination**

a term indicating the reaction of a fibre with chlorine when used with reference to textile processing. Chlorine can be in the form of a gas, or its solution in water or it may be obtained from a suitable compound.

■ **chogaichoga**

a garment worn over an inner garment, generally loose-sleeved coat-like garment like the angarakha (q.v.). It is sumptuous and appropriate for ceremonial occasions. Also known as Chauga, famous in Russia, Turkey.

■ **choli**

a short, bodice-like breast garment, with back covering or without tied with strings. It is of wide popularity famous among women in India.

■ **cholu**

a loose, shirt-like garment.

■ **chrome dye**

a mordant dye that helps in the formation of a chelate complex with a chromium atom.

■ **chrome mordant process**

process of dyeing the fibre is severed with a solution of a chromium compound and subsequently dyed with a suitable chrome dye.

■ **chromophore**

a part of the molecular structure that is responsible for colour. It has an organic dye or pigment.

■ **chrysalis**

the form taken by a silkworm in the dormant stage of development between larva and moth. It is dark brown in colour and fragments of it can often be detected in especially noils.

■ **churidar**

a 'pyjami' with bangle-like gathers or wrinkles, as in a churidar payan.

CIF
Cost, Insurance and Freight.

circular jersey
fabric produced on circular knitting machines.

circular knitting
made on a circular machine to produce tubular fabric such as jersey, sweaters, seamless hosiery and neckwear.

ciré
a lightweight performance fabric with a shiny surface made from synthetic fibres for use in outerwear. Fabrics made of thermoplastic fibres like nylon or polyester are cored by calendering with heat and pressure alone. Other fabrics like rayons or silks are calendared with wax or other compounds.

cisele velvet
a contrast formed by cut and uncut loops in velvet with a pattern.

class-four wool
these fibres are from 25-400 mm long, are coarse and hair like, have relatively few 'scales' and little crimp and therefore, more smoother and more lustrous. This wool is less desirable with the least elasticity and strength.

classification by fleece
wool shorn from young lambs differs in quality from that of older sheep. Also, fleeces differ according to whether they come from live or dead sheep, which necessitates standards for the classification of fleeces.

classing
a process of separating whole fleeces into different classes before being baled and sold.

class-one wool
merino sheep produce the best wool, which is relatively short, but the fibre is strong, fine and elastic and has good working properties. Merino fibre has the greatest amount of crimp of all wool fibres and has a maximum number of 'scales' two factors that contribute to its superior warmth and spinning properties. These sheep produce class one wool.

class-three wool
these fibres are about 100-455 mm long, are coarser and have fewer 'scales' and less crimp than merino and class-two wools. As a result, they are smoother and therefore, they have more lustre. These wools are less elastic and resilient. They are nevertheless of good quality to be used for clothing. This class of sheep originated in the United Kingdom.

class-two wool
class-two wools are not quite as good as the merino wool, but this variety is nevertheless very good quality wool. It is 50-200 mm in length, has a large number of 'scales' and has good working properties. This class of sheep originated in England, Scotland, Ireland and Wales.

clean content
the amount of clean, scoured wool remaining after removal of all vegetable and other foreign material.

clean wool

usually refers to scoured wool but occasionally it describes grease wool that has a minimum amount of vegetable matter.

clear finishing

usually, worsteds are not brushed, but closely sheared to give the fabric a clean face and crisp feel. This is called clear finishing.

clip

with angora goats, refers to the amount of hair removed from a single animal. One season's yield of wool.

clip dot / clip spot

a design effect created by the use of extra yarns which are woven into the fabric at a certain spot and are then allowed to float over the fabric to the next spot, on a woven fabric.

clock reel

a device for winding hanks of yarn. Some come with various kinds of counters. An image of a clock reel can be found at the Illinois state museum site.

cloqué

a compound or double fabric with a figured blister effect, produced by using yarns of different character or twist, which respond in different ways to finishing treatments.
The blister may be created by several different methods such as printing with caustic soda or other chemicals, by weaving together yarns under different tension or by weaving together yarns with different shrinkage properties.

closed shed

the shed in which weaving of cross thread tissues, some of the warp yarns are crossed over others. Such as gauze weave and leno weave.

cloth

a general term used for most textile fabrics. The term was originally applied to wool fabric suitable for clothing.

clothes moth

the larva of this moth eats wool and other protein fibres. Various articles on fibre pests can be found here.

clothing wool

wool under 1.5 inches in length and distinguished from combing wools by their shorter length. Principal properties include softness, crispiness and felting ability.

cloud yarn

a term given to yarns of irregular twist obtained by alternately holding one of the component threads while the other, being delivered quickly, is twisted around it and then reversing the position of the two threads, thus producing alternate clouds of the two colours.

cloudy wool
wool that is off-colour. It may be due to wool becoming wet while poorly stored in a pile.

cluny lace
a heavy bobbin lace in geometric patterns, done by using thick yarns usually of cotton or linen. Most often used for curtains doilies and trim for apparel.

coarse wool
wool that has a blood grade of 1/4 or common or a numerical count grade of 44's, 45's, or 48's, or micron count above 31. Coarse wool may have as few as 1 to 5 crimps per inch.

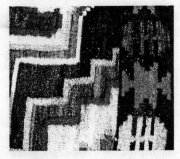

coated
refers to the application to make a fabric water repellent or waterproof. It's done of material such as plastic resin, wax, oil, varnish or lacquer to the surface of the fabric and the methods include dipping, spraying, brushing, calendaring or knife coating.

coated fleeces
some wool producers coat their fleeces that cut down on the amount of vegetable matter and weathering.

cockade
a ribbon bows generally a decorative piece deriving from the tie attaching the brim of a cocked hat. Often worn in contrast.

cocked hat
a hat of the styles of 17th or 18th century, which is styled with the brim, turned up.

cocoon (silk)
an oval-shaped casing done by silkworm to protect itself and it provides silk spun.

cocoon stripping
the first threads secreted by the silkworm when it finds a place to form its cocoon.

coiling head
a device that deposits the sliver in even coiled layers in tall cylindrical cans at the front of carding machines, drawing frames and combing machines.

coir
a reddish-brown-to-buff coloured coarse fibre obtained from the fruit of the palm cocos nucifera. This seed fibre is obtained from the husk of the coconut. Used in brush making, doormats, fish nets, cordage.

cold drawing (synthetic filaments and films)
the drawing of synthetic filaments or films without the intentional application of external heat. Note: free drawing of filaments or films at a neck is also referred to as cold drawing even though this may be

carried out in a heated environment.

1. sensation. The characteristic of the visual sensation that enables the eye to distinguish differences in its quality, such as may be caused by differences in the spectral distribution of the light rather than by differences in the spatial distribution or fluctuations with time.
2. of an object. the particular visual sensation (as defined above) caused by the light emitted by, transmitted through, or reflected from the object. Note: the colour of a non-self luminous object is dependent on the spectral composition of the incident light, the spectral reflectance or transmittance of the object and the spectral response of the observer. colour can be described approximately in terms of hue, saturation and lightness or specified numerically by chromaticity co-ordinates e.g., those defined by the C.I.E. standard observer data (1964). Alternatively, colour can be specified by reference to visual standards, e.g., the munsell colour atlas.

■ **collapse yarn**

collapse yarn is (usually) an over spun single, dried under tension that is then knit or woven. When the item is moistened, the yarn returns to its original elastic state.

■ **colour**

the actual colour of the wool. In industry a bright white to cream is most desirable, canary stains, brown or black stains are undesirable.

■ **colour constancy**

the ability of a coloured object to giving the similar colour effect when viewed under different illuminants, the observer having been chromatically adapted in each case.

■ **colour defect**

any colour that is not removable in wool scouring, due to urine stain, dung stain, canary yellow stain or black fibres.

■ **colour fast**

a term used to describe fabrics of sufficient colour retention so that no noticeable change in shade takes place during the normal life of the garment. Strictly speaking, no fabric is absolutely 'colourfast'. In buying fabrics make sure that they are fast to the particular colour hazard they will encounter.

■ **colour fastness**

that property of a dye, to retain its original hue, when handled under normal conditions when exposed to light, heat, or other conditions.

■ **colour quality**

a specification of colour in terms of both hue and saturation, but not luminance.

■ **colour schemes**

1. monochromatic, or one-colour harmony: the use of one colour in varying degrees of intensity and value light blue, medium blue, dark blue.
2. harmony: an agreeable combination of colours, all related to one another.

3. **complementary harmony:** a pleasing combination of complementary colours. One of the two could be used in larger areas than the other, the colours might show a difference in value and intensity and there should be no clashing.

■ **colour value, tinctorial value**
yielding of a colourant, compared with a standard of equal cost. It is usually determined by comparing the cost of colouration at equal visual strength. Comparisons are normally made between products of similar hue and properties.

■ **colour yield, tinctorial yield**
the depth of colour obtained when a given standard weight of colourant is applied to a substrate under specified conditions.

■ **comb (warper or slasher)**
a chain of upright metal teeths used for the separation of the individual warp strands and guides them onto a beam in proper order.

■ **combed fibres**
fibres that have combed. This process removes the short fibres. When drawn off, the fibre is called 'top'.

■ **combed yarn**
yarn produced from carded and combed fibres.

■ **comber sliver**
the loose, untwisted strand of cotton fibres produced with the help of combing machines from ribbon lap.

■ **combination yarn**
a yarn of fibre and filaments in which there are dissimilar component yarns.

■ **combing**
the combing process is an additional step beyond carding. In this process the fibres are arranged in a highly parallel form and additional short fibres are removed producing high quality yarns with excellent strength, fineness and uniformity.

■ **combing in oil**
the preparing and combing of wool to which oil has been added to facilitate the manipulation of the fibres.

■ **combing machine**
a machine that helps in removing short fibres, dirt and neps and straightening the remaining fibres into parallel alignment and prepares ribbon lap for spinning into fine yarn.

■ **combing wool**
wools having sufficient length and strength to comb. According to industry standards, the length of fibres for strictly fine combing must be over 2.75 in., with an increase in length as the wool becomes coarser.

■ **combing, dry**
the preparing and combing of wool to which no oil has been added.

■ **comforter**
an 'over-covering' on a bed that is made with a fabric shell filled with an insulating material.

commingled yarn
a yarn consisting of two or more individual yarns that have been combined, usually by means of air jets.

compact
a tight, thick fabric with a firm hand.

complements
these are colours that are opposite one another on the hue circle.

composite
a hard product, which consists of two or more discrete physical phases, including a binding material (matrix) and a fibrous material.

composite yarn
a yarn composed of both principle and consecutive-filament components, like core spun or wrap spun.

composite, composite material
a product formed by intimately combining two or more discrete physical phases-usually a solid matrix, such as a resin and a fibrous reinforcing component.

compressive shrinkage
a process in which fabric is shrinked in length e.g., by compression. Also known as controlled compressive shrinkage.

conch or conque
a large shell-shaped hat in gauze or light crepe, but generally much bigger and made of pale gauze, seems to have been of high fashion in England.

condense dye
a dye that reacts covalently with itself or other compounds during or after the application.

condenser card
a roller-and-clearer type of card, as distinct from a flat card, which converts fibrous raw materials slubbings, by means of a condenser. It divides a broad thin web of fibres into narrow strips, which then consolidated by rubbing into slubbings.

condenser spun
description of yarn spun from slubbing.

condensing
process of dividing the wide sheet of cotton fibres into a number of narrow ribbon-like strands which, when acted upon by the leather belts and rollers of a condenser, are formed into loose heavy strands (called roving) ready for spinning.

condition
in grease wool, the amount of yolk and foreign impurities it contains a fleece having a 'heavy condition' would have a large amount of shrinkage.

conditioner tube
a tube supplied with steam or hot air surrounding a melt-spun thread-line and located between extrusion and wind-up, whose purpose is to control the fine structure of the yarn., cone.

conditioning

the act of exposing bobbins to remove kinks from the yarn and to prevent its kinking in subsequent processes by filling yarn to steam or to a spray of conditioning solution in order to set the twist.

cone

1. a tapered cylinder of wood, metal or cardboard around which yarn is wound.
2. a package of yarn wound into a convenient shape.

consistency

the uniform distribution of all the fibre characteristics within each lock and throughout the entire fleece.

contemporary

currently in vogue..

continuous filament strand (glass)

a fibre bundle composed of many glass filaments.

continuous yarn felting

a process used to produce a yarn, or consolidate a spun yarn, whereby slivers, rovings, slubbings or yarns are felted on a continuous basis. Passing wool-rich material through a unit where it is distraughted by an aqueous medium where felting takes place does this.

continuous-filament yarn, filament yarn

a yarn made of one or more filaments known as monofilament or multifilament, that runs essentially the whole length of the yarn.

conversational

facetious designs or designs with a theme.

converter

a person or a company that buys grey goods and sells them as finished fabrics. A converter organises and manages the process of finishing the fabric to buyers' specifications, particularly the bleaching, dyeing, printing, etc.

converting, conversion (tow)

the production, which maintains the essential parallel arrangement of the filaments from a filament tow or tows. Note: the two methods of converting most commonly employed are:
1. crush cutting, in which the filaments of the tow are severed by crushing between an anvil roller and a cutting roller with raised 'blades' helically disposed around its surface.
2. stretch breaking, in which the filaments of the tow are broken by progressive stretch between successive, sets of rollers. If subsequently a top is required, further processes of re-breaking and/or gelling may be necessary and the whole operation is then often referred to as tow-to-top converting or conversion.

cool

a soft, smooth, hand generally associated with synthetics.

cool colours

colours that are light in nature like blue, violet and green are cool. They are reducing in nature, they

Textile

give a cool feel as seen by the eyes. Cool colours have a calm and restful effect.

■ **cooling cylinder**

an open or a closed cylinder filled with cold water, through which hot fabric is passed to accelerate cooling..

■ **cop**

a self-supporting package of yarn, which does not have a core through its centre. A form of yarn package spun on a mule spindle. The term can also be used to describe a ring tube.

■ **cop winding machine**

a machine, which winds yarn into cigar-shaped packages in small, headless, coreless.

■ **copolymer**

a polymer in which there are two or more repeat units.

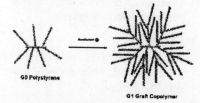

G0 Polystyrene
G1 Graft Copolymer

■ **copolymer, block**

a copolymer in which the repetition of units in the main chain occur in blocks.

■ **copolymer, graft**

a copolymer formed when sequences of one repeating unit are built as side branches onto a backbone polymer derived from another repeating unit.

■ **copp**

this refers to the cone of fibres that builds up on a spindle.

■ **cord**

a term used to describe the way in which textile strands have been twisted, such as in cabled or plied yarns. It includes:
1. cabled yarns.
2. plied yarns
3. in structures made by plaiting, braiding or knitting.

■ **corded**

1. a fabric with a surface rib effect resulting from the use of a heavier or plied yarn together with finer yarns.
2. a yarn made from two or more finer yarns twisted together.

■ **cordoban leather**

it is derived from goatskin, simply tanned, basically concerned the art of preparing this leather came from Cordoba.

■ **cordon yarn**

a two-ply union yarn manufactured from a single cotton yarn and a single worsted or woollen yarn.

■ **corduroy**

a cut filling-pile cloth with narrow to wide wale that runs in the warp direction of the goods and made possible by the use of an extra set of filling yarns in the construction. The back is of plain or twill weaves, the latter affording the better construction. Washable types are available and stretch and

durable press garments of corduroy are very popular. Usually an all-cotton cloth, some of the goods are now made with nylon or rayon pile effect on a cotton backing fabric or with polyester-cotton blends.

■ **core sampling**

a method of collecting representative samples from bales or packs of textile fibres obtained by inserting a coring tube driven by hand or machine into each package.
1. core samples can be used for the determination of yield or fineness, but not fibre length.
2. the term mini-core sampling is applied to small-scale sampling.

■ **core yarn**

a yarn made by winding one yarn around another to give the appearance of a yarn made solely of the outer yarn.

■ **core-spun yarns**

consist of a filament base yarn, with an exterior wrapping of loose fibre that has not been twisted into a yarn. Polyester filament is often wrapped with a cotton outer layer in order to provide the strength and resiliency of polyester, along with the moisture-absorbent aesthetics and dye affinity of cotton. Sewing thread as well as household and apparel fabrics is made from these yarns.

■ **core-testing**

the coring of bales or bags of wool to determine the clean content (or 'condition') and yield.

■ **core-twisted yarn**

a yarn produced by combining one fibre or filament with another during a twisting process.

■ **cornet**

the cornet headdress or a kind of cap with abridged frontage and upstanding frill in front and lappets at the back.

■ **correct invoice weight**

the weight of matter calculated from the oven-dry weight and the recommended allowance.

■ **cortex**

the inner portion consisting of spindle-shaped cells on most animal hair fibres.

■ **cortical cells**

the spindle shaped cells forming the inside structure of a fibre.

■ **cotted**

a fleece that contains fibres that are matted (or 'felted') together.

■ **cotton**

a cellulose fibre collected from the perennial shrub from the genus gossypium, predominantly hirsutum (upland or long-staple cotton), but also some g. barb dense (pima or extra-long-staple cotton). A vegetable fibre consisting of unicellular hairs attached to the seed of the cotton plant. Most cotton is coloured a light to dark cream and its chemical composition is almost pure cellulose. Coloured cottons in shades of tan, greens, blue and rust are also less commonly available. A distinct feature of the mature fibre is its spiralled or twisted.

■ **cotton count**

the cotton count expresses the number of hanks required to make a pound of yarn. A hank of cotton is equal to 840 yards. So 1 cc = 840 yards of cotton, the coarsest cotton yarn. A 3 cc yarn would then be one-third as course and would be expressed as 3/1 cc show that it is a single strand. Likewise plies are designated by two numbers separated by a slash such as 4/2 cc. This equals 3360 yards (4 x 840) of two-ply yarn. This yields 1680 yards of yarn per pound (3360/2). An 8/4 cc yarns would yield the same number of yards per pound, but would be a 4 plies of finer yarn. So a number 8 four-ply yarn is the same diameter as a number 4 two ply yarn.

■ **cotton dust**

dust which is present during the handling or processing of cotton that contains a mixture of substances, including smaller particles of ground-up plant matter, fibre, bacteria, fungi, soil, pesticides, non-cotton plant matter.

■ **cotton types**

1. acala: Mexican variety introduced into the U.S. medium staple cotton grown in the southwestern states and now in Israel.
2. American: upland cotton grown in this country. It forms bulk of world's crop. Fibre runs from 1/2" to 3/4".
3. American peeler: a variety of cotton grown in the Mississippi delta. Fibres range from 1 1/8 to 1 1/4 inches in length. Used in combed yarns and in fabrics, such as lawns, dimities and broadcloths.
4. American pima cotton: a cross between sea island and Egyptian. Grown in Arizona, brownish colour. Fine strong cotton. Averages 1 3/8" to 1 5/8". Used for sheer woven fabrics and fine knitted goods.
5. China: harsh, wiry, very short staple. Can be mixed with wool for blankets. Limited uses.
6. Egyptian: fine lustrous long staple cotton. Several varieties - usually brown in colour. 1 2/5" average. Used in U.S. for thread and fine fabrics.

7. Indian: cotton grown in India, for many years the consistent second largest cotton producing country. The Indian cottons imported are generally of the harsh, short staple type (8/10" to 9/10") for such uses as blanket filling.
8. Peruvian: a variety of cotton whose fibres average 1 1/4" in length and which comes from brazil, central America and the west Indies.
9. sea island: finest of all cotton, very white and silk-like with staple of 1.5 inches or better. Can be spun easily to 100s or better for exceptionally fine cloths. Before the war between the states was raised on the islands of the Carolinas and Georgia.

■ cotton waste

there are two different types of wastes known as 'hard' and 'soft' and their treatment differs according to the class. Hard waste comes from spinning frames, reeling and winding machines and all other waste of a thread nature whereas Soft waste is derived from earlier processes where the fibres are relatively little twisted, felted or compacted.

■ cotton wool

a web or batt of fibres used for medical or cosmetic purposes that is made from cotton and/or viscose rayon.

■ cotton yarn measures

54 inches = 1 thread (circumference of warp reel), 80 thread = 1 lea = 120 yds, 7 leas = 1 hank = 840 yards, 1 bundle = 10 pounds (usually).

■ cotton, long staple

cotton fibre of not less than 11/8 inches in actual staple length.

■ cotton-like

refers to a fabric that feels like cotton.

■ cotton-spun

a term applied to staple yarn developed from processing cotton into yarn produced on machinery originally developed.

■ cotty wool

wool that has matted or felted on the sheep's back. Caused by insufficient wool grease being produced by the sheep, usually due to breeding, injury or sickness. This type of defective wool is more common in the medium to coarse wools. The fibres cannot be separated without excessive breakage in manufacturing.

■ count

the number given to a yarn of any material, usually indicating the number of hanks per pound of that yarn. May also refer to the fineness to which a fleece may be spun. There are at least three definitions. In worsted yarn, the number of 560-yard skeins weighing one pound (Bradford method). In woollen yarn, the number of 256-yard skeins weighing one pound (yorkshire method).
1. the number of picks and warp ends per inch in cloth.
2. a number assigned to yarn to describe its fineness. The number is based upon number of hanks per pound of yarn.

count of cloth

the number of ends and picks per inch in a woven fabric as counted by an individual. If a cloth is 64 x 60, it means that there are 64 ends and 60 picks per inch in the fabric. A cloth that has the same number of ends and picks per inch in woven goods is called a square cloth, 64 x 64. Pick count is the term that is synonymous with texture or number of filling picks per inch.

countervailing duty

an extra duty imposed on an imported product by an importing country (or group of countries, as in the case of the EU) to compensate for subsidies deemed to be illegal which are given to the manufacturer of the product in the exporting country.

counting glass

a small mounted magnifying glass for examining fabric, which contains a unit of measuring. It does have an aperture one-centimetre square, one inch square or cross-shaped with various dimensions, convenient for counting ends and picks or courses and Wales in a fabric.

Count-strength product (CSP)

the product of the lea strength and the actual count of cotton yarn.

couple

to combine a suitable organic component to form an azo compound, usually a phenol or an arylamine, with a diazonium salt or in after treatment of direct dyeing.

course length (weft-knitted)

the length of yarn in a knitted course.

course, knitted (fabric)

a row of loops across the width of a fabric.

couvrechef

a veil or covering for the head.

cover factor (woven fabrics)

a number that indicates the extent to which the area of a fabric is covered by one set of threads. For any woven fabric, there are two cover factors a warp cover factor and a weft cover factor. Under the cotton system, the cover factor is the ratio of the number of threads per inch to the square root of the cotton yarn count.

covered yarn

yarn made by feeding one yarn through one or more revolving spindles carrying the other (wrapping) yarn. Covered yarn may also be produced using air-jet technology.

coverstock

a permeable fabric used in hygiene products to cover and contain an absorbent medium.

covert

a warp-faced fabric, usually of a twill weave, with a characteristic mottled appearance obtained by the use of a grandrelle (two-colour twisted yarn) or mock grandrelle warp. It is generally used in for over coating for both men and women. It is also made waterproof and used a great deal in rainwater.

■ **crabbing**

a term used in the textile industry. Crabbing sets the cloth and yarn twist by rotating the fabric over cylinders through a hot-water bath, or through a series of progressively hotter baths, followed by a cold-water bath. Crabbing is done to stabilise the fabric before dyeing and finishing and is necessary only for worsted fabrics.

■ **crank**

the extension of the 'axle' to the 'footman'.

■ **crash**

a coarse woven, very rugged and substantial in feel kind of fabric with a rough surface, made with thick uneven yarns. It comes in white or natural shades or could be dyed, printed, striped, or checked. It is strong, irregular in diameter, but smooth used for table linens, draperies, backings.

■ **cravat**

wide cloth or piece of lace knotted or tied around the neck.

■ **crease versus wrinkle**

a crease is a line or mark produced in anything by folding, a fold or furrow. A wrinkle is a ridge or furrow on a surface caused by contraction, folding, rumpling, etc. A crease is a deformation in a fabric intentionally formed by pressing. A wrinkle is formed unintentionally by washing and wearing and it can usually be removed by pressing. In permanent press, however, the crease is not removable and there is an absence of wrinkles.

■ **crease-recovery**

the measurement quantitatively in terms of crease-recovery angle specified of crease-resistance.

■ **crease-resist finish**

a finish, usually applied to fabrics made from cotton or other cellulose fibres or their blends, which improves the crease recovery and smooth-drying properties of a fabric. In the process used most commonly, the fabric is impregnated with a solution of a reagent, which penetrates the fibres and after drying and curing, cross-links the fibre structure under the influence of a catalyst and heat. The crease-resistant effect is durable to washing and to normal use.

■ **crease-resistant**

this refers to the ability of a fabric to resist creasing. Wool is considered to be very crease resistant, while cotton is not.

■ **creel**

1. a structure for holding supply packages in textile processing. This fibre characteristic may be expressed numerically as the crimp frequency or as the difference between the lengths of the straightened and crimped fibre, expressed as a percentage of the straightened length.
2. the rack for holding packages of roving or yarn on any textile machine.

3. the task of mounting packages of roving or yarn on the rack (creel) of any textile machine.

■ **crepe**

a fabric characterised by an all over crinkled, pebbly or puckered surface. The appearance may be a result of the use of high twist yarns, embossing, chemical treatment or a crepe weave. Has a crinkled, puckered surface or soft mossy finish. Comes in different weights and degrees of sheerness. Dull with a harsh dry feel and also woollen crepes are softer than worsted. If it is fine, it drapes well. Has very good wearing qualities and slimming effect uses, depending on weight, it is used for dresses of all types, including long dinner dresses, suits and coats.

■ **crepe de chine**

a very sheer silk crepe. They are of two types:
1. a fabric of medium lustre, woven from raw silk creepiness is obtained by degumming the fabric.
2. a more lustrous fabric made in Japan with a spun silk warp and a thrown silk filling. As now made, a sheer flat crepe in silk warp or manmade fibres. Used for lingerie, blouses and dresses.

■ **crepe georgette**

a sheer, dull-textured fabric, with a crepe surface, obtained by alternating left and right-handed yarns.

■ **crêpe yarn**

a highly twisted yarn which may be used in the production of crêpe fabrics.

■ **crepe-back satin**

a satin fabric in which highly twisted yarns are used in the filling direction. A two-faced fabric in which one side is crepe and the other satin. Also called satin-back crepe. The floating yarns are made with low twist and may be of either high or low lustre. If the crepe effect is the right side of the fabric, the fabric is called satin-back crepe.

■ **crepey**

the term refers to a fabric with a pebble like texture.

■ **crepon**

a fabric with a pleat-like crinkle effect in the warp (lengthwise) direction of the fabric, made with high twist yarns. Crepe effect appears in direction of the warp and achieved by alternates or slack, tension or different degrees of twist. It is mostly used in dresses and ensembles.

■ **crêpon**

a crêpe fabric that is more rugged than the usual crêpe with a fluted or crinkled effect in the warp direction.

■ **cretonne**

a plain weave fabric, usually cotton, with a neutral ground and brightly coloured floral designs, similar to chintz but with a dull finish and sometimes heavier. Used for draperies and upholstery. Some are warp printed and if they are, they are usually completely reversible. Designs run from the conservative to very wild and often completely cover the surface.

They are generally used in bedspreads, chairs, draperies, pillows, slipcovers, coverings of all kinds, beach wear, sportswear.

■ **crewel**

a type of embroidery using a loosely twisted 2 ply worsted yarn.

■ **crimp**

the wave effect in the wool fibre. Usually the finer wools show the most crimp. Uniformity of desired crimp generally indicates superior wool.

■ **crimp contraction**

the contraction in length of a previously textured yarn from the fully extended state (i.e. where the filaments are substantially straightened), owing to the formation of crimp in individual filaments under specified conditions of crimp development.

■ **crimp frequency**

the number of full waves or crimps in a length of fibre divided by the straightened length.

■ **crimp recovery**

the ability of a yarn or fibre to return to its original crimped state after being released from a tensile force.

■ **crimp retraction**

see **crimp contraction**.

■ **crimp stability**

the ability of a textured yarn to resist the reduction of its crimp by mechanical or thermal stress.

■ **crimp, latent**

a crimp that potentially exists in specially prepared fibres or filaments and that can be developed by a specific treatment such as thermal relaxation or tensioning and subsequent relaxation.

■ **crimped length**

the distance between the ends of a fibre measured with respect to its general axis of orientation, when substantially freed from external restraint.

■ **crinkle**

the waviness of each individual fibre when separated from a lock. It is responsible for elasticity and is usually irregular.

■ **crinkled**

an uneven, wrinkle, or puckered effect on the fabric surface, which can be created by a variety of mechanical or chemical, finishes, or through the use of high twist yarns.

■ **crinoline**

a heavily sized, stiff fabric used as a foundation to support the edge of a hem or puffed sleeve.

Also used as interlining, in the millinery and bookbinding trades and to give fullness to skirts. Usually comes in black, white or brown.

■ **crisp**

describes a fabric that has a smooth, soft and clean surface, good body and a relatively firm hand, which makes noise when rustled.

■ **Critical Application Value (CAV)**

the amount of finishing liquor with low add or easy care, which must be applied to a given fabric to avoid a non-uniform distribution of cross-linking after drying and curing.

■ **crocking**

the rubbing-off of dye from a fabric. Crocking can be the result of lack of penetration of the dyeing agent, the use of incorrect dyes or dyeing procedures or the lack of proper washing procedures and finishing treatments after the dyeing process.

■ **crockmeter**

an apparatus for evaluating the colourfastness to rubbing of dyed or printed textiles.

■ **croop**

silk, especially after immersion in a weak acid, when compressed and rubbed, gives out a peculiar rustling sound, which is known as 'croop'.

■ **cross cut**

refers to a corduroy fabric that has the pile cut in a weft wise direction, forming squares or rectangles on the surface.

■ **cross dyeing**

varied colour effects are obtained in the one dye-bath for a cloth that contains fibres with varying affinities for the dye used. For example, a blue dyestuff might give nylon 6 a dark blue shade, 6 a light blue shade and have no affinity for polyester thereby leaving the polyester area unscathed or white. It is a very popular method.

■ **cross lapping, cross laying**

the production of a non-woven web or batt from a fibre web by ranging it to and fro to the direction of traverse.

■ **crossbred wool**

a sheep bred from two distinct breeds, also a classification for wool of medium fineness. In the U.S., wool obtained from sheep of long-wool x fine-wool breeding. Usually, this wool grades at 3/8 or 1/2 blood.

■ **cross-linking**

the creation of chemical bonds between polymer molecules to form a three dimensional polymeric network, for example in a fibre or pigment binder.

■ **cross-wound package**

a package characterised by the large crossing angle of the helixes of sliver or yarn.

■ **crumbs**

a term used to describe shredded alkali-cellulose.

crush cutting

a process of converting in which crushing between anvils roller severs the filaments of the tow and a cutting roller with raised 'blades' helically disposed around its surface.

crushed

a finish that creates a planned irregular disturbance on the surface of the fabric, usually by mechanical means.

crutched wool

wool that has been clipped from rear end and udder area of ewes in the early spring to prevent collection of manure and fly strike.

culottes

french word for rather tight breeches.

cupra (fibre)

describes the fibres of regenerated cellulose obtained by the cuprammonium process.

cuprammonium

a process of producing a type of regenerated rayon fibre. In this process, the wood pulp or cotton liners are dissolved in an ammoniac copper oxide solution. Bamberg rayon is a type of cuprammonium rayon.

cupro

a type of cellulose fibre obtained by the cuprammonium process. A term used to describe fibres of regenerated cellulose obtained by the cuprammonium process.

curcuma

a fabric with a yellow colour similar to that produced by the curcuma spice.

curing (chemical finishing)

a process carried out after the application of a finish to a textile fabric in which appropriate conditions are used to affect a chemical reaction. Usually, the fabric is heat treated for several minutes. However, it may be subject to higher temperatures for short times (flash curing) or to low temperatures for longer periods and at higher regain (moist curing).

curl yarn

a type of yarn, which presents curls or loops of various sizes all along its surface. It is usually produced as follows two threads, a thick and a thin are twisted together, the thin being held tightly and the thick thread slackly twisted around it. This two-fold yarn is then twisted in the reverse direction with another thin thread, this untwisting throwing up the thick thread as a loop, the two fine threads holding the loops firmly.

cut

a length of warp or a woven cloth required to weave a piece of cloth.

cut and sew

a system of manufacturing in which shaped pieces are cut from a layer of fabric and stitched together to form garments. In the case of tubular

cut velvet | de-aeration

knitted fabric, the cloth is either cut down one side and opened up into a flat fabric or left as a tube and cut to shape.

■ **cut velvet**

beaded velvet or jacquard fabric consisting of a velvet design on a plain ground and used in evening wear and home furnishings.

■ **cuticle**

the outer layer of cells of a fibre which are hard, flattened and do not fit together evenly and whose tips point away from the fibre shaft forming serrated edges. These serrated edges cause the fibres to grip together during processing and manufacturing.

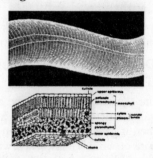

■ **damask**

firm, glossy jacquard-patterned fabric first brought to the western world by Marco polo in the 13th century. Damascus was the centre of fabric trade between the east and west, hence the name. Damask is similar to brocade but flatter and reversible. It may be linen, cotton, rayon or silk, or combination of fibres. Used for tablecloths, napkins, draperies and upholstery.

■ **damp wool**

wool that has become damp or wet before or after bagging and may mildew. This weakens the fibres and seriously affects the spinning properties.

■ **deacetylated acetate (fibre) (generic name)**

a term that describes the process by which fibres of regenerated cellulose obtained by almost completes de-ethanoylation (deacetylation) of a cellulose ethanoate (acetate).

■ **dead cotton**

an extreme form of immature cotton with a very thin fibre wall, commonly the cause is excessively slow secondary growth, resulting in many of the fibres having developed only a thin secondary wall by the time the boll opens. It is sometimes caused by premature 'death' or cessation of growth due to factors such as local pest attack.

■ **dead wool**

wool taken from the sheep that have died on the range or have been killed. Wool recovered from sheep that have been dead for some time is occasionally referred to as 'merrin'. Wool taken from sheep that have died on the range or have been killed. Dead wool fibre is decidedly inferior in grade and is used in low-quality cloth.

■ **de-aeration**

separating all undisclosed gases and part of the dissolved gases (chiefly air) from solutions prior to extrusion.

■ deburring
a process used for extracting burrs, seeds and vegetable matter from wool. A burring machine carries out this process mechanically.

■ decitex
a unit of weight indicating the fineness of yarns and equal to a yarn weighting one gram per each 10,000 metres. The abbreviation for this is 'd'tex'.

■ Decitex Per Filament (DPF)
the average decitex of each filament in a multifilament yarn.

■ deco
designs suggesting the art deco style of the 20's and 30's, classified by bold outlines and streamlined shapes.

■ decortication (flax)
the process of removing woody outer layers from the stem of the flax plant to yield flax fibres.

■ decrystallised cotton
cotton when treated with reagents to reduce the degree of crystallisation. It is done with zinc chloride, concentrated caustic soda solutions or amines.

■ deep dyeing
1. impact made of fibres modified so as to have greater uptake of selected dyes than normal fibres, when the two are dyed together.
2. the removal of grease, suint and extraneous matter from wool by an aqueous or solvent process.
3. the removal of natural fats, waxes, grease, oil and dirt from any textile material by extraction with an organic solvent, degree of orientation, the extent to which the macromolecules composing a fibre or film lie in a predominant direction in the case of fibres the predominant direction is usually the fibre axis. Note a): there are several methods for assessment of the degree of orientation, of which measurement of birefringence is the most usual. b): the degrees of orientation of crystalline and non-crystalline regions may be evaluated separately.

■ defective wool
wool that contains excessive vegetable matter, such as burs, seeds and straw or which is campy, clotty, tender or otherwise faulty.

■ degreasing
any method that removes yolk and dirt from wool.

■ Degree of Polymerisation (DP)
the repeating of units in the individual macromolecules in a polymer, it generally, depends on the basis on which it is calculated, which should be stated. For example, it may be based upon a mass (weight) or a number average.

■ degummed silk
the process of removing gum (sericin) from the yarn/fabric by boiling the silk in hot water. This way the lustre of the silk is enhanced. It is very lightweight.

■ degumming
the boiling-off of silk in silk and hot water, in order to dissolve and

wash away the natural gum (seracin) which surrounds the fibre.

■ **delaine wool**

fine, strictly combing wool, usually from Ohio and Pennsylvania. Delaine wool does not necessarily have to come from the delaine-merino; however, that breed is noted for this class of wool.

■ **délavé**

a fabric with a washed effect.

■ **delicate**

referring to a fine, light hand with good drape.

■ **delocalisation**

the geographical move of a production unit to a low cost country. (Note that the term is increasingly being used to describe all forms of shifts in production, including foreign sourcing and subcontracting.).

■ **delustrant**

a particulate material added before extrusion to subdue the lustre of a man-made fibre.
1. the anatase form of titanium dioxide is commonly used for this purpose.
2. terms used to indicate the level of delustrant in man-made fibres include: clear, bright, semi-dull, semi-matt, dull, matt, extra dull and super dull.

■ **demi-lustre wool**

wool that has some lustre but not enough to be classed as lustre wool. Wool of this type is produced by the romney and similar breeds.

■ **denier**

a system of measuring the weight of a continuous filament fibre. In the United States, this measurement is used to number all manufactured fibres (both filament and staple) and silk, but excluding glass fibre. The lower the number, the finer the fibre, the higher the number, the heavier the fibre. Numerically, a denier is the equivalent to the weight in grams of 9,000 metres of continuous filament fibre.

■ **denim**

this staple cotton cloth is rugged and serviceable and is recognised by left-hand twill on the face. Coarse single yarns are used most, but some of the cloth used for dress goods may be better quality stock. A two-up and one-down or a three-up and one-down twill may be used in the weave formation. Standard denim is made with indigo-blue-dyed warp yarn and a grey or mottled-white filling. It is the most important fabric in the work-clothes group and it is used for overalls, wear field and has even been used as eveningwear. Popular also in the upholstery and furniture trades. Its uses includes work clothes, overalls, caps, uniforms, bedspreads, slipcovers, draperies, upholstery, sportswear, of all kinds, dresses and has even been used for evening wear.

■ **density**

an index of the number of wool fibres per unit of a sheep's body. Fine-wool breeds show greater fleece density than the coarser wool breeds.

■ **dents/inch**
a unit of measure that denotes the number of reed wires and spaces between adjacent wires in one inch.

■ **depitching**
the removal branding substances from wool, usually, by solvent-extraction.

■ **depth**
that colour quality which increases with an increase in the quantity of colourant present, all other conditions (viewing, etc.) remaining the same.

■ **design draft**
a diagram depicting the method that is to be woven into a cloth and also the basic weave (plain, twill or satin) of the cloth to be produced.

■ **design paper**
cross-section paper on which design drafts are made.

■ **desizing**
the removal of size from fabric.

■ **detergent**
a substance specifically intended to cleanse a substrate normally having surface-active properties.

■ **detwisted**
descriptive of a yarn of fibres or filaments from which twist has been removed.

■ **devantière**
17th century women's riding costume split at the back.

■ **developed dyes**
used on cotton and rayon and other manmade fibres, on the latter when developed from disperse dye bases. Light-fastness rated from poor to good.

■ **developing**
a step in a colouring or printing process in which an intermediary form of the colorant is converted to the final form (e.g. oxidation of a vat leuco ester).

■ **devoré**
the production of a pattern on a fabric by printing it with a substance that destroys one or more of the fibre types present.

■ **dhila**
assorted or baggy. Thus, a *dhila payjama*, wide and roomy all over.

■ **dhoti**
the customary Indian dress for the lower part of the body, consolidating of a piece of unstitched cloth draped over the hips and legs and can be worn in various ways in different parts of the country, alike by men and women.

■ **diacetate (fibre)**
a term used to describe fibres made from propanone-soluble (acetone-soluble) cellulose ethanoate (acetate). The ISO generic name is acetate.

■ **diamond**
referring to designs dominated by diamond shapes.

diazotise
to convert a primary perfumed amine into the matching diazonium salt, by treatment with nitric acid.

die swell
the upsurge in diameter that occurs as a visco-elastic melt or solution that appears from a die or spinneret hole.

differential dyeing
usually evocative of fibres of the same primitive class, but having potentially different dyeing properties from the standard fibre.

diffusion
progress of substance owing to the existence of a concentration gradient.

dimity
a plain weave with a crosswise or lengthwise spaced rib or crossbar effect, done on a cotton fibre. A thin fine with corded spaced stripes that could be classified in single, double or triple. It has a crisp texture, which remains fair even after washing and resembles lawn in the white state. It is easy to sew and manipulate and launders well, easily mercerised and has a soft lustre.
Uses: children's dresses, women's dresses and blouses, infant's wear, collar and cuff sets, bassinets, bedspreads, curtains, underwear. Has a very young look.

dingy
wool that is dark or greyish in colour and generally heavy in shrinkage. May be caused by excessive yolk, poor farming conditions or parasites.

diolen
a high tenacity polyester filament yarn produced by acordis.

dip
1. a fascination of a textile in liquid for relatively short duration.
2. the intensity of liquid in the inner cylinder of a rotating washing machine.
3. a laboratory dyeing, usually to develop a dye formula, (U.S.A.).

dip dyeing
a process in which a garment is dipped into a dye bath to achieve dye take-up only in those areas immersed.

dip-dyed yarns
yarns produced by dip dyeing.

direct dyes
a class of aniline dyes, so called because they have such great affinity for cellulose fibres, i.e., cotton and linen. While both these and acid dyes are sodium salts of dye acids, direct dyes do not require the use of a mordant. Their shades are duller than those of either acid or basic dyes and they tend to have less tincture value than the basic dyes. However, they have the very important advantages of being much more lightfast than the basic dyes and possibly more so than acid dyes. It includes:
1. (man-made fibre production) integrated polymerisation and fibre extrusion without intervening isolation or storage of the polymer,

2. (man-made fibre production) the method whereby tow. It is converted to staple fibre and spun into yarn in an integrated operation.
3. (bast fibre production) a method of dry-spinning bast fibres whereby untwisted slivers are drafted with suitable controls and directly twisted into yarn. Gill spinning and slip-draft spinning systems are particular forms of the method.

■ **direct printing**

also known as roller, calendar or cylinder printing, the colours are printed directly onto the fabric in the same manner as the printing of wallpaper or a newspaper. There must be one roller for each colour used and some machines can handle as many as sixteen colours. Bleached goods are fed into the machine and pass between the colour rollers and the master or main cylinder. The colour rollers are etched, each with the respective part of the entire motif that it will supply to make up the completed design. Most direct prints have a white background or base. Chintz and cretonne are good examples of direct printing.

■ **direct style**

a pattern of printing one or several colours where the dyes are applied and then fixed by ageing or other appropriate means.

■ **direct warping**

the direct conversion of yarn from a package basket on to a beam.

■ **direction of twist**

to determine twist, hold yarn in a vertical position and examine the angle of the spiral. The angle of the s twist will correspond to the centre portion of the s. the angle of the z twist will correspond to the centre portion of the z. When spinning, the wheel should rotate counter clockwise for an s twist and rotate clockwise for a z twist.

■ **direct-spun**

1. a term used to describe strings or yarns produced by direct gyrating.
2. evocative of woollen yarns slant on a mule on to weft bobbins.

■ **dirty tips**

the weathering that occurs on the ends of some locks. These may not completely wash out or evenly dye.

■ **discharge or extract printing**

used to obtain medium to dark coloured fabrics with white or coloured motifs. The cloth is first of all piece dyed and colour is then discharged or bleached-out in certain areas leaving them white. The cloth is then direct-printed and some or all of the white areas are coloured to provide the finished effect.

■ **discharge printed**

a tinted fabric is printed with a chemical paste that peroxides or 'discharges' the colour to allow white patterns on a dyed ground. It doesn't get affected by adding a dye to the paste. The best possible way is to replace the discharged ground colour with another colour.

■ **discharging**

the obliteration by chemical means of a dye or acerbic, already existing on a material to leave a white or differently coloured pattern. The term also refers to cover the removal of gum from silk (see **degumming**).

■ **disperse dye**

significantly water-insoluble dye having substantivity for one or more hydrophobic fibres. Some of the major examples are cellulose acetate and usually applied from fine aqueous dispersion.

■ **dispersion spinning**

a process in which the polymers be inclined to an infusible, insoluble and generally stubborn character e.g., polytetrafluoroeth-ylene. They are diffused as fine particles in a carrier such as sodium alginate or sodium xanthate solutions that allows extrusion into fibres, after which they are caused to coalesce by a heating process, the carrier being removed either by a heating or by a dissolving process.

■ **dissolving pulp**

a specially distilled form of cellulose made from wood tissue.

■ **distaff**

a staff with a cleft or formed-end for holding flax from which the fibre is drawn in spinning. May be attached to a spinning wheel. The monastic heritage museum shows a wheel-mounted distaff on a wheel from the 1800s.

■ **distribution layer**

a layer in a non-woven hygiene product (such as a diaper), which distributes fluid to a super absorbent and/or fluff pulp material, where it is absorbed.

■ **district check**

distinctive woollen checks originally made in different districts of Scotland. A category of small check designs, sometimes with contrasting over plaids.

■ **diz**

the small tool that is used to help form and even top in wool combing. Traditionally a diz was made out of carved horn. You can also make (or buy) very nice ones out of wood. A cheap, non-e-classy alternative is to trim a piece of plastic and punch or drill a hole in the middle of it. This is done with the bottom corners of a plastic milk jug or a crescent cut from a section of PVC pipe. As always, if it involves worsted spinning, please.

■ **DMT**

Dimethyl Terephthalate-a chemical intermediate used in the manufacture of polyester.

■ **dobby**

a general term for a fabric woven on a special dobby loom, which allows the weaving of small, geometric figures. A dobby weave can often be distinguished from a plain weave by the patterns are beyond the range of simple looms.

dobby loom

a type of loom on which small, geometric figures can be woven in as a regular pattern. Originally this type of loom needed a 'dobby boy' who sat on the top of the loom and drew up warp threads to form a pattern. Now the weaving is done entirely by machine. This loom differs from a plain loom in that it may have up to thirty-two harnesses and a pattern chain.

Automatic Dobby Dress Goods Loom

dobby weaves

a decorative weave, characterised by small figures, usually geometric, that are woven into the fabric structure. Dobbies may be of any weight or compactness, with yarns ranging from very fine to coarse and fluffy. Standard dobby fabrics are usually flat and relatively fine or sheer. However, some heavyweight dobby fabrics are available for home furnishings and for heavy apparel.

doeskin

properly a leather made from the skin of the doe. Also used to describe:
1. a heavy five or eight-shaft satin-weave cotton fabric napped on one side.
2. a heavy short-napped woollen fabric used for men's wear. The term doeskin finish should be used. Its uses include women's suits and coats and also in a lighter weight for dresses. Sportswear and riding habits for both men and women. Trousers and waistcoats for men.

doff

to remove, a filled package or beam from a textile machine. Frequently the operation includes replacing the full package or beam with an empty one, as in doffing a drum carder.

doffing comb

a fluctuating jagged steel bar, which strips the cotton from the doffing cylinder in a light film or sheet when set adjacent to the doffing cylinder of a carding machine.

doffing cylinder

a drum covered with wire-tooth on a carding machine that tiles the cotton in a light film from the carding drum, which is exposed in turn by the doffing comb.

doffing tube (rotor spinning)

an expansion done to the navel to guide the reserved yarn from the rotor.

doggy

wools that have no character and show the results of lack of breeding. These wools are usually short, coarse and lacking in feel.

dogstooth or houndstooth check

a small colour and weave effect using a $^2/_2$ twill.

dolly

1. a machine in which fabric pieces are sewn end to end and are disseminated repeatedly through a liquor by means of a single pair of squeeze rollers above the liquor.
2. a machine in which lace, hosiery, or knitwears are made subject to the action of free-falling beaters when engrossed in a detergent solution and approved in a moving rectangular or cylindrical box.
3. an open-width washer, which contains around 3-5 compartments, originally used for fertilising aged cotton prints.

domestic wools

all wools grown in your own country as opposed to those imported.

domestics

1. ordinarily cotton goods such as unbleached muslin sheeting or print cloth.
2. general term to cover household fabrics such as blankets, sheeting and pillow casing, towels, washcloths.
3. American-made carpets and rugs as distinguished from those made in other countries, especially those known as oriental rugs.
4. fabrics made in this country from the major textile fibres as distinguished from those made in the British isles or on the continent - covert, serge, cassimere, Melton, kersey, beaver, broadcloth, organdy, voile, dimity and others.

domett flannel

generally cotton woven plain and twill and in white. Has a longer nap than on flannelette. It consists of soft filling yarns of medium or lightweight are used to obtain the nap. The term is identical with 'outing flannel' but it is only made in a plain weave. Both are soft and fleecy and won't infuriate the skin and any sizing or starching must be removed before using. It is mostly used for infants' wear, interlinings, polished cloths.

donegal

originally homespun woven, plain-weave fabric from woollen-spun yarns characterised by a random distribution of brightly coloured flecks or slubs. It was originally produced as a coarse woollen suiting in county donegal. Commonly used in coats, heavy suits, sportswear. Has a tailored, sporty look.

donegal tweed

a medium to heavy of plain or twill weaves fabric in which colourful yarn slobs are woven into the fabric. The name originally applied to a hand-woven woollen tweed fabric made in Donegal, Ireland. End-uses include winter coats and suits.

dope

a spinning solution of fibre-forming polymer as primed for extrusion through a spinneret. A spinning solution is often referred to as dope, a term historically associated with cellulose ethanoate (cellulose acetate) solutions as varnishes.

dope/solution dyed

contrived fibres, which are coloured by dyeing the polymer solution before it is protruded and spun into yarn.

doru

a lengthy rope with which the thick woollen coat worn by the gaddis is held around the waist.

dosuti

'two threads', it literally means in Hindi, used to describe the operation of combining two threads together at a meandering machine and because of this the operation is known as 'dosuti winding'.

dot

a design subjugated by circular spots of any size, printed or woven into the fabric. In it small dots are often called pin dots, whereas medium to large dots may be referred to as aspirin dots, coin dots or polka dots.

dotted Swiss

sheer cotton fabric embellished with small dot motifs. The dots may vary in colour and can be applied to the goods by swivel weaving, clip spotting, flock-dotting, lappet weaving, etc. Originated in saint gallen, Switzerland, in about 1750. Uses include dress goods, curtaining, eveningwear, wedding apparel, baby clothes, etc.

double (yarn)

also termed plied yarn.

double cloth

a fabric construction, in which two fabrics are woven on the loom at the same time, one on top of the other. In the weaving process, the two layers of woven fabric are held together using binder threads. The woven patterns in each layer of fabric can be similar or completely different.

double coated

some breeds of sheep (and other fleece-bearing animals) have two coats. Sometimes the double-coating refers to different colours, perhaps a dark outer/longer coat. Sometimes this refers to the length. Also referred to as 'primitive'.

double drive

belts drive both the flyer and bobbin from the drive wheel. The bobbin pulley (or whorl) is smaller, which determines the spinning ratio. Some double drive wheels can be converted to run with scotch tension.

double face

a changeable fabric consisting of 2 layers, usually with a different colour or pattern on every side. Double face is usually a double cloth but some reversible bonded fabrics may be referred to as double face.

Textile double fleece | doupion, douppioni

- **double fleece**

  a fleece consisting of two year's growth.

- **double knit**

  a fabric knitted with a double stitch on a double needle frame to provide a double thickness and is the same on both sides. It has excellent body and stability.

- **double weave**

  a woven fabric construction made by interlacing two or more sets of warp yarns with two or more sets of filling yarns. The most common double weave fabrics are made using a total of either four or five sets of yarns.

- **double-face satin**

  yarn that are woven with two warps and one filling, to replicate a double satin interpretation. It has satin on both sides, but in cheaper qualities, cotton filling is often used.

- **doubling**

  1. the process of joining two or more strands of roving or sliver and drawing out the resulting strand. The cause behind this operation is to increase the consistency of the cotton strand and ultimately of the yarn made from it.
  2. the act of meandering two or more elements of yarn onto one package without twisting them.

- **doubling machine**

  a machine that creases cloth to half or quarter of its original width.

- **doublings (drawing)**

  the number of laps, rovings, slivers or slubbings, noshed simultaneously into a machine for drafting into a single end. It is engaged to promote blending and regularity.

- **doup**

  a special kind of heddle, used in juxtaposition with ordinary heddles on the harnesses of a loom to cross and uncross warp filaments in both a horizontal and vertical plane.

- **doupion, douppioni**

  silk yarns derived out of the cocoon of two silk worms that have nested together. In spinning, the double strand is not separated so the yarn is uneven and irregular with a large diameter in places. Fabric is of silk made in a plain weave and is very irregular and shows many slubs - seems to be through in a hit and miss manner. It is reproduced in rayon and some synthetics and one such famous fabric is called 'cupioni'.

down twist

this is one of the two terms that expert use when discussing plying. This refers to 's-twist'. The expert maintains that people get so hung up trying to remember whether an s-twist is spun clockwise, that they lose track of process. It really doesn't matter whether your singles are spun s or z, you just need to ply them in the opposite direction.

down wool

also called 'hill wool'. Wool of medium fineness produced by such breeds as the Southdown and the Shropshire. These sheep are distinguished by their fine and curly wool of short staple, which is especially adapted for making loose, rough, moss-like, felted, carded yarns for the production of clothing. These wools are lofty and well suited for woollen. Much of the down wool runs $1/4$ to $3/8$ blood in quality. This can be a great wool for felting.

downproof

a fabric that resists the diffusion of down. It may be closely woven to be downproof by nature or may be cired or encrusted to make it down proof.

draft

1. see **design draft**.
2. to draw out (attenuate or stretch) a strand of cotton, usually by running the strand between several pairs of rollers, each pair turning faster than the pair before it.

drafting

a process, which reduces the linear density of an assembly of fibres. Drafting typically occurs in the early stages of producing yarns from staple fibres.

drafting triangle

the small triangle of fibres that are formed between your drafting hand and your fibre hand. This should never be longer than the fibre length. Also called a 'drafting triangle'.

drainage (geotextiles)

the ability of a geotextile to collect and transport fluids. Liquids or gases are transmitted within the plane of the geotextile and this involves flow across the geotextile. For example, geotextiles are used to capture and transmit gases (e.g. methane) beneath the geomembrane in a landfill capping system.

drape

the way a fabric hangs. Drape is affected by yarns, weave structure and finish.

drapery

decorative fabrics for the home. Made of cotton, silk, rayon, nylon, spun glass, acrylic, polyester, wool, mohair and mixtures. Used as:
1. hangings at the side of windows and doors for artistic effects,
2. inner curtains of sheer materials to hang next to the windwindowpane. Draw curtains to insure privacy or to shut out light as sash curtains.

■ **drapey**

refers to a fabric with good swathe, one that is supple and falls easily into graceful folds when suspended or tailored.

■ **draping**

draping means to lynch or to embellish the body form with loose fabric and to attain the best out of waste, i.e. a body fitted garment by using adequate sewing techniques.

■ **draw (mule)**

the series of operations from the start of the external run to the finish of the internal part of the horse and carriage of a spinning or a twiner mule.

■ **draw (sampling)**

a sample of fibres preoccupied manually from a bulk lot of raw material. The term is also referred with sliver with a view to deem the length and/or distribution of length of fibre within the sample.

■ **draw mechanism (knitting)**

an apparatus on a straight-bar knitting machine for exchanging rotary motion into responding motion for the purpose of laying the yarn and fading it round the needles.

■ **draw out**

see **draft**.

■ **draw pin**

a stationary pin or guide, which by suggesting a localised change in yarn rigidity and/or temperature may be used to stabilise the position of the draw-point or neck in some processes of drawing of man-made-fibre yarns.

■ **draw ratio**

machine draw ratio calculated from drawing process, the ratio of the peripheral speed of the draw roller to that of the feed roller. They are of 3 types:
1. true draw ratio, in a drawing process, the ratio of the linear density of the undrawn yarn to that of the drawn yarn.
2. residual draw ratio, the draw ratio required, in draw texturing, to convert a partially oriented yarn into a commercially acceptable product.
3. natural draw ratio, the ratio of the cross-sectional areas of a filament before and after the neck, when a synthetic filament or film draws at a neck.

■ **draw roller**

the production roller of a zone in which drawing takes place.

■ **draw sliver**

the assorted, raveled strand of cotton fibres that is the product of the drawing frames.

- **draw spinning**

  a process for spinning partially or highly oriented filaments in which the orientation is introduced after melt spinning but prior to the first forwarding or collecting device.

- **draw texturing**

  a process in which the drawing stage of synthetic yarn manufacture is combined with the texturing process.

- **draw threads (lace)**

  detachable threads that are included in the building of lace either acting as a temporary support for certain parts of the pattern or to hold together narrow widths or units that are separated subsequently by their removal.

- **draw twist**

  a process of orienting a filament yarn by drawing it and then twisting it in integrated sequential stages.

- **draw-beaming**

  also termed warp drawing.

- **draw-down**

  its a artificial filament extrusion, the ratio of taking-up or carry off speed to the average speed of the spinning fluid as it leaves the spinneret.

- **drawing (staple yarn)**

  operations by which silvers are merged or doubled or levelled and by drafting reduced to the state of sliver or roving suitable for spinning.

- **drawing (synthetic filaments and films)**

  the extending to near the limit of plastic flow of synthetic filaments or films of low molecular orientation. This process adjusts the molecular chains in the length direction.

- **drawing frame**

  a machine in which the processes of merging of several strands of sliver are into one strand and drawing out takes place, so that the combined strands approximate the weight and size of any one of the original strands.

- **drawing in**

  the process of threading the warp filaments from a beam through the heddles and reed of a loom in the order designated on a design draft.

- **drawing roll cleaner**

  a pad of felt or similar material, which assists to remove the dust and lint that collects on the drawing rollers as they sketch out the roving or sliver. It is attached to the underside of the cleaner box cover.

- **drawing rollers**

  two or more pairs of rollers, each pair of which turns at a higher speed than the previous pair. It helps in serving to draw out or assuage the roving or sliver passing between them.

- **drawing, hot (synthetic filaments and films)**

  a term applied to the drawing of synthetic filaments or films with the intentional application of external heat. Free drawing of filaments or films at a neck is also referred to as cold drawing even though this may be carried out in a heated environment.

■ **drawing-in**

the process of drawing the threads of a warp through the eyes of a healed and the dents of a reed.

■ **drawing-in frame**

a frame for holding a beam of warp strands, harnesses and reeds to make assure that the strands are drawn easily through the harnesses and reeds in a specified order.

■ **drawing-in hook**

a tool similar to a crochet hook, used to draw the individual warp filaments through the heddles and reed of a loom.

■ **drawn yarn**

extruded yarn that has been subjected to a stretching or drawing process that orients the long-chain molecules of which it is composed in the direction of the filament axis. On further stretching, such yarn acquires elastic extension as compared with the plastic flow of undrawn yarn.

■ **draw-spinning**

a process for gyrating partly or highly oriented filaments in which the orientation is introduced prior to the first forwarding or collecting device.

■ **draw-twist**

to orient a filament yarn by drawing it and then by twisting it in integrated sequential stages.

■ **draw-warping**

a process for the preparation of warp beams or section beams from a creel of packages of moderately adjusted yarn in which the conventionally separate stages of drawing and beaming are combined sequentially on one machine.

■ **draw-wind**

to familiarise a filament yarn by drawing it and then to wind it on to a package in an incorporated process without imparting twist.

■ **dress muslin**

see **muslin**.

■ **dressing (lace)**

the operation done to the stretching lace, net, or lace-furnishing products to size, then drying, after the application of stiffening or softening agents. These processes of stretching and drying may be carried out on either a running stenter or a stationary frame.

■ **drill**

a twill fabric, usually piece-dyed, similar in construction to denim. Some are left in grey but can be bleached or dyed. When dyed a khaki colour, it is known by that name. Uses: uniforms, work clothes, slipcovers, sportswear and many industrial uses.

■ **drip-dry**

vivid of textile materials that are plausibly resistant to interruption of fabric structure and appearance during wear and washing and require a minimum of ironing or pressing.

■ drive band

the cord that runs between the wheel and the flyer. A single-drive band is a circle and is used with the scotch tension wheels. A double-drive band is a figure-8 folded back on itself and loops over the flyer unit and the speed whorl.

■ drive ratio

ratio of wheel diameter to flyer whorl diameter (or bobbin whorl on a bobbin lead wheel). Governs how much twist you get in the yarn for each treadle. To measure your wheel ratios, set up your wheel, tie a bright-coloured piece of yarn to your flyer arm and adjust the treadle until it is at the bottom of its movement. Slowly rotate the wheel, while counting the flyer revolutions until the treadle returns to its original point. The bright yarn tied to the flyer arm just makes it easier to count.

■ drop spindle

a spindle that hangs freely from the fibre source (as opposed to a supported spindle). Probably so named by people who haven't added enough twist.

■ drop stitch

knit fabrics under this caption are constructed to control the degree of not looping certain stitches and to provide for opening needle latches when necessary. The drop stitch construction is generally limited to jersey and rib fabrics for either fabric design or for the separation of rib fabric pieces. Used in knit shirts and dress fabrics.

■ drop wire

a flat piece of metal, with a hole in it, for the warp filament to pass. It drops and stops the machine when the filament threaded through it breaks.

■ drum carder

a rotating drum, covered with carding cloth, used to card fibres. An example is hand-cranked drum carder.

Ashford Drum Carder
available with either fine or coarse teeth

■ dry

refers to a fabric that is lacking in surface moisture or natural lubrication and needs to be worked upon.

■ dry cleaning

a cleansing method or process applied to garments in which organic solvents such as carbon tetrachloride, perchloroethylene or certain hydrocarbon compounds are used to remove dirt, soil and most spots and stains. Unaffected stains have to be removed by other special agents.

■ **dry combing**

preparing wool for worsted spinning without any oil. Also referred to as 'French combing'.

■ **dry laying**

a method of forming a fibre web or batt by labelling, followed by a bonding process.

■ **dry spinning**

in the dry spinning process, polymer is dissolved in a solvent before being spun into warm air where the solvent evaporates. This leaves the fibrous polymer ready for drawing.

■ **dry-combed top**

a wool top including not more than 1 % of fatty matter based on the oven-dry, fat-free weight as tested by the international wool textile organisation's method.

■ **drying cylinder**

heated, revolving, hollow cylinder(s) around which textile material or paper is passed in contact with it.

■ **dry-laid**

part of a production route for making non-woven, in which a web of fibres is produced either by carding or by blowing the fibres on to an endless belt.

■ **dry-spun**

1. descriptive of a bested yarn made from a dry-combed top or of synthetic yarns spun on similar machinery.
2. expressive of coarse flax yarn spun from air-dry roving (cf. wet-spun).
3. explanatory of man-made filaments produced by dry-spinning.

■ **dry-spun flax**

this is a term for spinning flax and mainly is a way of differentiating it from 'wet-spun flax'. In dry-spun flax, additional water is not added to the surface in spinning. It produces a hairier, less-attractive yarn.

■ **duchess**

this form of satin has a wonderful lustre and a smooth feel. It is a dress fabric with 8-12 shaft fabric. The material is string, has a high lustre and texture and it is firm in nature. Usually 36"in width.

■ **duchesse lace**

a guipure lace characterised by floral and leaf designs with very little ground. heavier threads are intertwined to give raised texture. They are used in bridal veils and gowns.

■ **duck**

the name duck covers a wide range of fabrics. It is the most durable fabric made. A closely woven, heavy material. The most important fabrics in this group are known as number duck, army duck and flat or ounce duck. Number and army ducks are always of plain weave with medium or heavy ply yarns, army ducks are the lighter. Ounce ducks always have single warp yarns woven in pairs and single or ply-filling yarns. Generally of ply yarns in warp and yarns of various sizes and weights in filling.

■ **dull**

a yarn or fibre surface lacking in lustre. Descriptive of textile materials, the lustre of which has been reduced.

■ **dumping**

the offer for sale of large quantities of goods in a foreign market at low prices, usually in order to gain market share, while maintaining higher prices in the home market. Dumping may be deemed to have taken place when a product is sold in a foreign market at a price, which is less than the cost of production plus a normal profit margin.

■ **dupaluidupallari top**

minor, skin-tightening cap made generally of muslin and consisting of two similar pieces cut slightly rounded and curved towards the top.

■ **dupatta**

veil-cloth worn by women, wrapped lightly around the upper part of the body.

■ **dupion**

a silk-breeding term meaning double-cocoon. Hence, an irregular, raw, rough silk reeled from double cocoons.

■ **dupion fabric.**

originally a kind of silk fabric woven from doupion yarns. The term is nowadays functional to imitations woven from man-made-fibre yarns.

■ **duplex printing**

also called register printing, it is the printing of both sides of the goods with the same or different motifs. Woven design effects are often simulated in this work. Curtains, hangings, some upholstery fabric and some sportswear are printed by this method.

■ **durability**

the ability of a fabric to resist wear through continual use.

■ **durable finish**

any type of finish rationally resilient to normal usage, washing and/or dry-cleaning.

■ **durable press**

a finishing treatment designed to impart to a textile material or garment the retention of specific contours, including defined creases and pleats, which are resistant to normal usage, washing and dry cleaning.

■ **dusting**

the second step in commercial wool processing (after sorting). The purpose is to remove as much dirt and sand as is possible before scouring.

■ **DWR (fabrics)**

Durable Water Repellent. DWR fabrics retain their ability to repel water after washing, dry cleaning or heavy wear.

■ **dye**

there are many application classes of dyes, including acid dyes, disperse dyes, reactive dyes and natural dyes. Dyes may be generally divided into natural and synthetic types. Natural dyes are obtained from berries, flowers, roots, bark and more. Synthetic dyes are chemical compounds.

■ **dye ability**

the capacity of fibres to accept dyes.

■ **dye bath**

the solution (usually water) containing the dyes, dyeing assistants and any other ingredients necessary for dyeing.

■ **dyed & overprinted**

refers to fabrics, which have been first dyed and then printed in different colours that are darker than the dyed base.

■ **dyed in the wool**

fabrics or yarns where the fibres were dyed prior to processing.

■ **dye-fixing agent**

a substance, generally organic, practical to a tinted or printed material to improve its fastness to wet treatments.

■ **dyeing**

the process of applying a comparatively permanent colour to fibre, yarn or fabric by immersing in a bath of dye.

■ **dyeing of textiles**

the process of applying colour to fibre stock, yarn or fabric, there may or may not be thorough penetration of colour into the fibres or yarns.

■ **eastern pulled wool**

wool is pulled from the skins after it has been loosened, usually be a depilatory. Pulled wool should not be confused with dead wool.

■ **easy care**

refers to fabrics, which are refurbished to their original appearance after laundering with little or no ironing. In general such fabrics can be machine-washed and tumble dried.

■ **ecru (knitting)**

descriptive of fibres, yarns or fabrics that they shouldn't be subject to processes affecting their natural colour.

■ **edge roll**

the usual curl or roll that develops at the edges of single knit fabric, making it rather difficult to handle since it does not lie completely flat.

effect threads
yarns supplemented in a fabric that adequately differing fibre, count or formation or enhance a pattern.

e-glass
a formulation of glass designed for use in electric circuitry, which has particularly good electrical and heat resistance properties. E-glass is also the most common type of glass formulation used in glass-fibre reinforcements.

Egyptian cotton
cotton from Egypt characterised by its strong, fine, long and glossy fibres.

elastane (fibre) (generic name)
a term used to describe fibres that are composed of synthetic linear macromolecules having in the chain at least 85% (by mass) of segmented polyurethane groups and which hastily revert to their original length after extension to three times that length.

elastic
a rubber band, cord, fabric or thread which has springiness, flexibility and resiliency. There are several types used in the textile trade today latex, cut rubber yarn, extruded latex, filatex, rolled latex and laton. These yarns are used in belts, garters, girdles, gloves, shoes, sportswear, suspenders, etc.

elastic fabric
a fabric that contains rubber or other elastomeric fibres or threads, having reparable extensibility in a direction parallel to the elastomeric threads and characterised by a high resistance to deformation and a high capacity to recover its normal size and shape.

elasticity
the ability to return to its original length after being stretched or compressed. Wool has more elasticity than cotton, with finer wools stretching up to 30% of their original length.

elastodiene (fibre) (generic name)
a term used to describe fibres composed of natural or synthetic polyisoprene, or unruffled of one or more dienes polymerised with or without one or more vinyl monomers, which quickly reverts significantly to its original length after extension to three times that length.

elastomer
a polymer, which has high extensibility together with rapid and substantially, completes elastic recovery.

elastomeric yarn
a yarn fashioned from an elastomer.
1. elastomeric yarn may either be incorporated into fabric in the

bare state or wrapped with relatively inextensible fibres. Wrapping is done by covering core spinning or uptwisting.
2. examples are elastane and elastodiene yarns.

■ **electret**

a non-conductive polymeric material, which can maintain a long-lived electrostatic charge. Polypropylene electret's filtration fabrics conveniently combine the mechanical removal of particles with an electrostatic field, which materially increases the filtration efficiency.

■ **electric 'wheel'**

a flyer on bearings driven by a small motor. Very compact, can be useful for people with limited use of legs. Very portable, can be battery powered.

■ **electrostatic flocking**

the process of affecting a herd to an adhesive-coated substrate in a high-voltage electrostatic field.

■ **elongation**

the ability of fibres in yarns or in fabrics 'to go in the direction of the weave'. Also means the increase in length from a tensile force example of this, the fibres in yarns as they appear in 'baggy trousers' on a rainy day or the sagginess of some woollen fabrics of low texture or pick count.

■ **elysian**

a thick, hard, heavy, usually woolen fabric with a deep nap that forms a slanting or undulate pattern on the surface. Used for coatings.

■ **emboss**

to manufacture a pattern in relief by passing fabric through a calendar. The process includes passing through from a heated metal bowl stamped with the pattern works against a relatively soft bowl, built up of compressed paper or cotton on a metal centre.

■ **embossed**

fabric with an improved design that has been firstly carved on a metal cylinder then dazed on the fabric with heat and pressure.

■ **embossing**

a popular effect made on cloth by passing it between a series of rollers, each set having one smooth and one embossed roller. These metallic rollers are heated so as to give better results. The embossed rollers have been engraved with suitable patterns, which will be reproduced on the fabric and give the appearance of a raised or embossed surface to the goods. Motifs may be birds, 'tear drops', foliage, scrollwork, figures, pastoral scenes, etc.

■ **embroidered**

a fabric or a kind of designing decorated with needlework stitching of yarn or thread. It can be done by hand or machine.

■ **embroidery**
a decorative pattern superimposed on an existing fabric by machine stitching or hand needlework.

■ **emerised**
a fabric, which has been passed over a series of emery-covered rollers to produce a suede-like finish.

■ **emerising**
a process in which the passing of a fabric through a series of emery-covered rollers is done, to produce a suede-like finish. A similar process is known as sueding.

■ **end**
1. (spinning) an individual strand.
2. (weaving) an individual warp thread.
3. (fabric) a length of finished fabric less than a customary unit (piece) in length.
4. (finishing) (a) each passage of a length of fabric through a machine, for example, in jig-dyeing. (b) a joint between pieces of fabric due, for example, to damage or short lengths in weaving or damage in bleaching, dyeing or finishing.

■ **end & end**
a plain weave fabric with a warp yarn of one colour blinking with a warp yarn of white or any other second colour. It is used most commonly in shirtings.

■ **end-group**
a chemical group that develops the end of a polymer chain.

■ **ends down**
1. a condition in which one or more ends have broken in a textile machine.
2. a defect in cloth that occurs when weaving is continued after ends have broken, without first mending them.

■ **engagéantes**
lace manacles with two or three rows of rumples, finishing women's gown sleeves.

■ **english combs**
the multi-pitch (commonly 4-pitch) hand combs used in preparing top.

■ **enthalpy**
the amount of energy in joules required to heat 1 gram of fabric from a temperature of 20°c to its melting point.

■ **entrepôt**
a trading centre or port at a geographically convenient location where goods are imported and re-exported without directly entering the local economy. According to the strict definition, goods are imported into and re-exported from an entrepôt without incurring liability for duties.

■ **enzyme washed**
refers to the process of cleansing with a cellulase enzyme -one, which attacks the cellulose in the fabric-giving it a second-hand, worn appearance and an attractive soft hand. The effect is similar to stone washing but is less damaging to the fabric. It is also known as bio washing. It is generally done with

denim or other cottons and fabrics of lyocell.

- **epidermal barrier**
  a barrier of the outer layer of skin.

- **epithelial tissue**
  a newly formed tissue.

- **epitropic fibre**
  a fibre whose surface contains incompletely or totally entrenched particles that adapt one or more of its properties. The best example is electrical conductivity.

- **eponge (souffle)**
  derived from the French term eponge for 'spongy'. A kind of fibre that is very yielding and sponge-like in a variety of fresh effects with loose weave of about 20 x 20. It is also known as ratine in cotton. Many stores now call eponge 'boucle'.
  It is mostly used in suits, dresses, coats, sportswear and summer suits.

- **eri**
  a sort of wild silk.

- **etamine**
  a fine wool crêpe.

ETAMINE TR 1

- **ethnic**
  refers to designs with elements symptomatic of the culture or customary designs of a peculiar group/groups of people.

- **evenness**
  this term refers to the uniformity of the fibre throughout the fleece.

- **exfoliation**
  an intrinsic fault in silk only, which is evident after degumming or dyeing. It is characterised by fine fibrils or fibrillae that is detached from the filament, giving it a speckled, dishevelled appearance.

- **exhaust treatment**
  a batch wise treatment in which a substance (such as a finish) is selectively adsorbed by a textile material immersed in the treatment liquor.

- **exhaustion**
  the amount of dye taken from the dye bath by the fibre, yarn or fabric being dyed. also, the condition of the dyer at the end of the day.

- **expression (per cent)**
  the weight of liquid reserved by textile material after the process of mashing or hydroextraction, premeditated as a percentage of the air-dry weight of the goods.

- **extension**
  fluctuation in length. The increase can be denoted in three ways, namely:
  1. on the basis of length.
  2. on the basis of percentage of the initial length.

3. on the basis of fraction of the initial length.

■ **extract**

wool or hair derived through the wet process of carbonisation.

■ **extrusion**

in the procedure of spinning of man-made filaments, fibre-forming substances in the plastic or dissolved state are strained through the holes of a spinneret or dye at a given rate. There are generally five methods of spinning (extruding) man-made filaments, but mishmash of these methods can also be used.

■ **eyelash**

a fabric with abrupt yarn on the facade suggesting eyelashes.

■ **eyelet**

a small hole or perforation made in series formation to receive a string or tape. It is worked around with a buttonhole stitch. Applied especially to garments make of broadcloth, dimity, organdie, piqué, etc.

■ **eyelet plate**

a kind of crossbar attached to an end of a creel exactly in front of each row of spindles. It is punched with the same number of holes, as there are spindles in the row and works to guide the individual ends from the packages on the spindles to the warping machine.

■ **fabric**

a woven, knitted, plaited, braided, or felt material such as cloth, lace, hosiery, etc. includes materials used in the manufacture. Fabric is also known as cloth, material, goods, or stuff. Garments are made from fabrics. There are three general classes of fabrics-apparel, decorative and industrial. Fabrics are most commonly woven or knitted, but the term includes assemblies produced by lace making, tufting, felting, net making and the so-called non-woven processes.

■ **fabric length**

generally the usable length of a piece between any fact marks, piece-ends or numbering, especially when the measurement is not specified and when the fabric is measured laid flat on a table in the absence of tension.

■ **fabric width**

until it is not specified exceptionally, it is the distance from one edge to another edge of a fabric when laid flat on a table without tension. Fabric is conditioned in a standard atmosphere for testing, in the case of commercial dispute. While buying and selling fabric one should specify the basis on which the width is assessed. For example overall, within limits or usable width (which implies within stenter pin marks).

■ **face**

the right side or the better-looking side of the fabric.

■ **face-finished fabric**

cloths finished only on the face. Much resorted to on melton's, kerseys and on other over coatings. The weaves used permit the type of finish, not withstanding the fact that the texture is high and the interlacing tight. Plain, twill and satin weaves are all used jointly in proper construction of the various face-finish cloths. Other face-finish clothes are Bolivia, bouclé, chinchilla, Montana, tree bark cloths, Saxony over coating and Whitney finishes.

■ **face-to-face carpets**

carpets produced as a cram in which the pile is attached alternately to two substrates. For example two cut pile cutting the pile yarns between the two substrates finishes carpets.

■ **facing silk**

a fine lustrous fabric of silk generally of a corded satin, twill weave or barathea, which is used for facing. For e.g., lapels in men's evening wear.

■ **facings**

binding of fine fur or intense cloth, these are a type of trimmings that are done merely for decoration. During the flow of time the meaning kept changing towards the contemporary meaning and the term now explains the covering of all the reveres of the body or sleeves of a garment.

■ **faconne**

a fabric with small scattered motifs usually jacquard but sometimes burn out. Faconne, in French, means fancy weave. Has small designs all over the fabric. Fairly light in weight and could be slightly creped. Background is much more sheer than the designs, therefore the designs seem to stand out. Very effective when worn over a different colour. Drapes, handle and wear well.

■ **faconne velvet**

attractive velvet made by burnt-out print process. The design is on velvet with a plain background.

■ **fad**

fleeting fashion are called fad's, they hardly have any ever-lasting impact on future fashion. They are briefly and suddenly visible for a short span and then they disappear.

■ **fade**

1. in fortress testing, any change caused by light or contaminants in the atmosphere in the colour of an object. Like, burnt-gas fumes. The change in colour could be in tone, intensity or brightness or any combination of these.
2. casually it can be said as a reduction in the depth of colour of an object, irrespective of cause. (a) straight, pinned bars employed in the control of fibres between drafting rollers and, (b) curved arms fixed to two shafts on a mule carriage and carrying the faller wires.

■ **faggoting**

an openwork hoop effect with connection of threads across the open area that creates a ladder effect.

■ **faille**

a ribbed silk or rayon cloth with crosswise rib effect. It is soft in feel and belongs to the grosgrain family of cross-rib materials. Used for coats, dress goods, handbags. Faille is rather difficult to launder well, has good draping effects and will give good service if handled carefully. Finished at 36 to 40 inches wide. Most commonly made with filament yarns but can be from a variety of fibres and weights. It usually has a soft hand and a light lustre with good body and drape. Its uses include dresses, blouses, soft evening purses and some dressy coats.

■ **faille crepe**

has a smooth, dull and richer face effect than crepe de chine. Fibre content must be declared if not made of all silk.

■ **faille taffeta**

made on plain weave, occasionally on twill construction, it is crisp and stiff and has a very fine cross-rib filling effect. Made in silk, rayon, or acetate that is used for coats and dresses.

■ **fairisle**

a type of sweater knitted with a coloured pattern in a traditional design originating in Scotland.

■ **fake furs**

cotton and manmade fibres are used in these woven or knitted fabrics, which have periodic waves of popularity. Their effects may be conservative or bizarre. Simulations of the fur of animals such as broadtail, chinchilla, ermine, French poodle, giraffe, krimmer, mole and pony are all well done. Bizarre effects are sometimes used in exaggerated markings and fanciful colourings. Of course, fabrics and garments made of fake fur do not have the actual warmth, generally speaking, of genuine fur but the articles may be dry cleaned and made flameproof. Fake furs find use in lounging apparel, dress and sports clothes, slippers, coats and jackets.

■ **fall wool**

wool shorn in the fall following 5-6 months of growth.

■ **falling bands**

a type of linen or lace collars, also known as rabat and hanging collars, with two distinct ends hanging down over the chest.

■ **false twist**

the major process used in texturising filament yarns. A rotating spindle twists the yarn, which is then set in a heater-box or tube, after which it is untwisted. Called by this name since the twist inserted does not become permanent. The twist, however, does remain in part because of the so-called 'memory' of the twist that was inserted in the processing. As a result, the yarn gains torque (the movement of forces that cause rotation or twisting as in the instance of twisting cord, wire, or yarn), or stretch, as well as bulkiness. To remove stretch the yarn is subjected to a second heat treatment, which affords stabilisation, but at the same time, retains bulkiness.

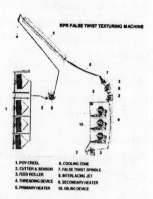

RPR FALSE TWIST TEXTURING MACHINE

1. POY CREEL
2. CUTTER & SENSOR
3. FEED ROLLER
4. THREADING DEVICE
5. PRIMARY HEATER
6. COOLING ZONE
7. FALSE TWIST SPINDLE
8. INTERLACING JET
9. SECONDARY HEATER
10. OILING DEVICE

■ **false-twist direction**

the twisting direction, say s or z, generated by a false-twisting device.

■ **false-twist texturing**

a process in which a single filament yarn is twisted, set and untwisted. When yarns made from thermoplastic materials are heat-set in a twisted condition, the deformation of the filaments is 'memorised' and the yarn is given greater bulk.

■ **false-twist-textured yarn**

an unremitting process, in which a yarn is extremely twisted, heat-set and then again untwisted. In an occasionally used different method, two yarns are continuously folded together, heat-set, then separated by unfolding.

■ **fancy yarn**

a yarn that differs from the normal creation of one and/or more folded yarns by way of deliberately produced irregularities in its construction. These irregularities are related to an increased input of one or more of its components or to the insertion of periodic effects such as knots, loops, curls, slubs or the like.

■ **fargul**

a kind of jacket.

■ **farji**

a kind of jacket. It was simply defined as 'a kind of garment', the farji was possibly a long over-garment without sleeves or with very short sleeves, open in front and worn like a coat over pyjama or angarakha.

■ **farshi pyjama**

wide-legged, loose from the legs. Simply saying, 'pyjama' that tracks on the ground, sometimes completely covering the feet, worn often with a kurta or angarakha.

■ **fasciated yarn**

a fasten fibre yarn that consists of a core of effectively parallel fibres compelled together by wrapper fibres by virtue of their manufacturing technique. The current technique referred to as jet spinning.

■ **fashion forecast**

predicting foretell future fashion tread for an exact period of time.

■ **fast colour**

a dye, which is stable to colour destroying agents, such as sunlight, perspiration, washing, abrasion and pressing.

■ **fastness**

possessions of resistance to an agency designated. For e.g., washing, light, rubbing, crocking, gas-fumes, etc. Noted points include: on the standard scale, five grades are usually recognised and are hideously changed from 5, demonstrating impassive, to 1. Different grades are used like for light fastness, eight grades are used, 8 representing the highest degree of fastness.

■ fatuhi
a type of 'jacket without sleeves'. By and large a vest lightly padded with cotton wool and quilted.

■ faux fur
a mound fabric made to replicate animal fur. Woven or knit in a variety of fibres although the most common among them are acrylic and modacrylic.

■ faux leather
a fabric made to impersonate animal leather. For e.g. polyurethane laminate.

■ faux linen
a fabric prepared from slubbed yarns to imitate linen. It is usually economical, easy care fabrics.

■ faux silk
a fabric of manufactured fibre, most commonly polyester, with good drape, lustre and a soft hand to imitate silk.

■ faz-vi
a 'jacket without sleeves'. Possibly the same kind of garment as fatuhi.

■ FDY
Fully Drawn Yarn.

■ feed roller, feed roll
a roller that forwards a yarn to a successive processing or take-up stage.

■ fell (of the cloth)
the edge of the fabric in a weaving looms formed by the last weft thread.

■ fellmongedring
the process of gathering wool from the fleeces of dead sheep.

■ felt
from the Anglo-Saxon meaning to felt or filter, a defecating device. The cloth is a matted, compact woollen material, of which Melton might be cited as an example. There are two types of felt cloth- woven and unwoven. Woven felt is what is concerned with here. The term may be misconstrued easily and not understood. Felting is another form of the word when speaking of cloth being 'felted'. Felting of woven cloth is perfected by an interlocking of the natural, scaly serrations on the surface of the contiguous wool fibres through the agencies of heat, moisture, steam, pressure and hammering. Some felted cloths have admixtures of hair fibres by agglutination. Many types of over coatings are correctly and incorrectly alluded to as being 'felt'. Uses can be categorised on the basis of many industrial uses, such as: piano hammers and in the printing industry. Many novelties, such as: pennants, slippers, lining of many kinds, insoles and toys. Hats and felt skirts.

■ feltability
the degree to which fibres will consolidate by felting.

felting

the rugging together of fibres during processing.

felting property

the property of wool and some other fibres to interlock with each other to create felt. Felting is caused by the directional friction effect of scales on the fibre surfaces. The factors involved in felting are the fibre structure, the crimp of the fibres, the ease of deformation of the fibre and the fibre's power of recovery from deformation.

FLBCs

Flexible Intermediate Bulk Containers. Large polypropylene woven containers used for packaging and carrying granulated bulk goods. FIBC are suitable for containing or carrying loads between 500 and 2,000 kg.

fibre

the fundamental unit comprising a textile raw material such as cotton, wool, etc. Fibres may be elongated single celled seed hairs like cotton, elongated multi cellular structures such as wool, an aggregation of elongated cells like flax, or manmade filaments like nylon, polyester, rayon. Fibre originally meant spinnable material including the natural fibres and short sections of manmade filaments. Such fibres have a length that is many times as great as their diameter. In order to be spun into a yarn, a fibre must possess sufficient length, strength, pliability and cohesiveness.

fibre fineness

the mean fibre diameter, which is usually, expressed in microns.

fibre length

the staple length of the fibre. On combing wools, this is often 3-8 inches, on the down wools 1.5-3 inches. With cotton, it may be 1/4-1 inch long. Bast fibres, like flax, may have a staple length of 36 inches.

fibre thickness

the average diameter of the fibre.

fibrefill

specially engineered manufactured fibres, which are used as filler material in pillows, mattresses, mattress pads, sleeping bags, comforters, quilts and outerwear.

fibrillation

the longitudinal splitting of a fibre or filament to give either micro-fine surface hairs or a complete breakdown into sub-micron fibres. In fabrics for apparel, fibrillation can be used to create a variety of surface textures and attractive aesthetics. In hydro entangled non-woven fabrics, the fibrils make entanglement easier and can give added strength to the fabric.

■ **fibroin**
a tough, elastic protein, which forms the principal component of raw silk.

■ **filament**
a fibre of indefinite or extreme length, some of them miles long. Silk is a natural filament, while nylon and polyester are synthetic filaments. Filament fibres are generally made into yarn without the spinning operation required of shorter fibres, such as wool and cotton. Filament yarns are smoother and more lustrous than spun yarns.

■ **filling**
in a woven fabric, the yarns that run cross the fabric from selvage to selvage and which run perpendicular to the warp or lengthwise yarns. Also referred to as the weft.

■ **filter cake (geotextiles)**
the graded soil structure developed upstream of the bridging particles on a geotextile acting as a filter.

■ **filtration (geotextiles)**
the process of retaining soil particles by a geotextile while allowing the passage of water. The geotextile allows the water and finer soil particles to pass through while retaining those of a coarser nature. A filter cake builds up on the face of the geotextile and this is where the actual filtration of the soil particles occurs. In order to perform this function a geotextile must be able to convey a certain amount of water across the plane of the geotextile throughout its design life.

■ **findings**
refers to pocketing, linings, zippers and other sundry and supplementary fabrics used in the manufacture of all types of garments.

■ **fine wool**
the finest grade of wool: 64's or finer, according to the numerical count grade or wool with an 18 to 24 micron count. Also, the wool from any of the Merino breeds of sheep. Fine wools may have as many as 30 crimps per inch.

■ **finish oil**
oil that is put on a yarn, either flat or textured, to reduce friction during subsequent processing stages.

■ **finished fabric**
a fabric that has gone through all the necessary finishing processes and is ready to be used in the manufacturing of garments.

■ **finishing**
this refers to additional steps used after the yarn is removed from the bobbin.

■ **fire retardant**
fabrics treated with special chemical agents to make them retardant or resistant to fire.

■ **fisheye**
large diamond-effect linen cloth that is similar in shape to the eye of a fish. Comparable with the smaller pattern noted in birds eye

and used for the same purposes. Durable, has food absorptive properties and is reversible.

■ **flame resistant**

a term used to describe a fabric that burns very slowly or has the ability to self-extinguish upon the removal of an external flame.

■ **flame retardant**

a chemical applied to a fabric or incorporated into the fibre at the time of production, which significantly reduces a fabric's flammability.

■ **flammability**

the ability of a textile to burn under specified test conditions.

■ **flammé**

a slob yarn.

■ **flannel**

a medium-weight, plain or twill weave fabric that is typically made from cotton, a cotton blend or wool. The fabric has a very soft hand, brushed on both sides to lift the fibre ends out of the base fabric and create a soft, fuzzy surface. End-uses include shirts and pyjamas.

■ **flannelette**

a medium-weight, plain weave fabric with a soft hand, usually made from cotton. The fabric is usually brushed only on one side and is lighter weight than flannel. End-uses include shirts and pyjamas.

■ **flash spun bonding**

a major variant of spun bonding, where polypropylene is solvent-dissolved and then pumped through holes into a chamber. The solvent is then flashed off and highly oriented filaments are produced.

■ **flash-spun**

a type of web made by flash spun bonding.

■ **flat crepe**

a firm silk or synthetic fibred fabric with a soft, almost imperceptible crinkled texture. Used for blouses, lingerie, dresses and linings.

■ **flat outerwear fabric**

this is made by having the needles arranged in a straight line, as distinguished from fabric made on a circular machine. Used in the manufacture of blouses, ensembles, scarves, skirting and sweaters.

■ **flat underwear fabric**

this is made on a machine with only one set of needles, to be distinguished from ribbed fabric, which requires two sets of needles.

■ **flax**

a slender, erect, annual plant (genus Linum) having narrow, lance-shaped leaves and blue flowers, cultivated for its fibre and seeds. The fibre of this plant, manufactured into linen yarn for thread in woven fabrics.

■ **fleece**

the woollen shorn from any sheep, or from any animal in the wool category. Fleece wool means clipped wool. As contrasted with pulled wool. Also the name of a fabric that has a deep fleece-like napped surface that may be wool, cotton, acrylic, nylon or other manmade fibres. Used for heavy coats.

■ **fleece lined**

a double-knit fabric, which floats on either on or both sides. These floats are napped, which makes the fabric warmer than an ordinary fabric. Used in elder down and cotton 'sweat-shirts'. Term also applied to sheep-lined coats.

■ **flicker**

a hand tool that looks like a small hand card on a long handle. To use it, hold one end of a lock of wool in your left hand rested on your thigh and 'flick' the tool up and down with your other hand catching the end of the fibre. This will open out the lock and make it easier to spin. It is recommended to wear a sturdy pair of jeans or place a leather pad on your left thigh.

■ **float**

the portion of a warp or filling yarn that extends over two or more adjacent warp ends or filling picks in weaving in order to form certain motif effects. Some floats are rather 'long' from the interlacing points with the opposite system of yarn.

■ **floating (warp)**

a length of warp yarn, which passes over two or more weft threads (rather than intersecting with them) in a woven structure.

■ **floating (weft)**

a length of weft yarn, which passes over two or more, warp threads (rather than intersecting with them) in a woven structure.

■ **flock**

a material obtained by reducing textile fibres to fragments by, for example, cutting, tearing or grinding.

■ **flock printing**

a process in which a fabric is printed with an adhesive, followed by the application of finely chopped fibres over the whole surface of the fabric by means of dusting-on, an air blast or electrostatic attraction. The fibres adhere to the printed areas and are removed from the unprinted areas by mechanical action.

- **flocking**

  a type of raised decoration applied to the surface of a fabric in which an adhesive is printed on the fabric in a specific pattern and then finely chopped fibres are applied by means of dusting, air-brushing, or electrostatic charges. The fibres adhere only to the areas where the adhesive has been applied and the excess fibres are removed by mechanical means.

- **flounce**

  hanging strips of material, which is normally sewn to the hem of a skirt.

- **fluorescent fabrics**

  made from dyestuffs that impart brightness to fabrics in daylight and under so-called 'black-light conditions'. This iridescent effect is used on apparel worn at night such as by firemen, policemen, airport workers, etc. Also effective for road signs.

- **flyer**

  a rotating device that adds twist to the roving and winds the stock onto a spindle or bobbin in a uniform manner.

- **flyer bearings**

  holds the flyer, same material as used in wheel bearings.

- **flyer lead**

  a single band drives the flyer. The bobbin has an adjustable friction band to slow it.

- **foam printing**

  a process in which a rubber solution is turned into a foam and squeezed through a screen to make a rubber print. Also known as puff rubber printing.

- **foamback**

  term used in Great Britain to denote that a fabric has been laminated to a backing of polyurethane foam.

- **FOB**

  Free-On-Board goods are delivered on-board a ship or to another carrier at no cost to the buyer.

- **folded yarn**

  a yarn made by twisting two or more single yarns together in one operation.

Folded yarns
another term for plied yarns.

follicle
the skin structure from which hair or wool fibre grows.

footman
the vertical connection between the treadle and the crank.

foulard
a lightweight twill-weave fabric, made from filament yarns like silk, acetate, polyester, with a small all-over print pattern on a solid background. The fabric is often used in men's ties.

FOY
Fully Oriented Yarn.

frame wheel
the flyer is usually mounted above the wheel, which means less floor space is used. Also called a 'castle wheel'. A well-known example of this is the Reeves Castle Wheel.

free swell absorbency
the weight of fluid in grams that can be absorbed by 1 gram of fibre, yarn or fabric.

free wools
usually means wool that is free from defects, such as vegetable matter.

French combing wool
wools that are intermediate in length between strictly combing and clothing. French combs can handle fine wools from 1.252.5 inches in length. Yarns that have been produced with the French combing method are combed dry, without oil added.

fribby wool
wool containing an excessive amount of second cuts and/or sweat locks.

friction angle (geotextiles)
an angle, the tangent of which is equal to the ratio of the friction force per unit area and the normal stress between the two materials and quantifies soil geotextile friction.

frisé
a fine bouclé yarn.

frowzy wool
a waste, lifeless-appearing, dry, harsh wool, lacking in character.

fugitive colours
dyes that fade, especially those that lose colour relatively quickly when exposed to natural light.

fuji silk
a spun-silk fabric woven in a plain weave.

Blush　Pearl Grey　Ivory

■ **full bleached**

this implies that the material has received at least on boiling in the alkali bath or baths and that bleaching has taken place in the bleaching bath.

■ **fulling**

the operation of shrinking and felting a woollen fabric to make it thicker and denser. Also called 'milling'. You can also full woollen yarn to give you a lovely knitting yarn. (If you were going to weave with the same yarn, you wait until you had woven the fabric.).

■ **fulling agent**

a chemical, usually a surfactant that acts as a lubricant during the process of fulling.

■ **gabardine**

firm, durable, compactly woven cloth, which shows a decided diagonal line on the face of the goods, made on a 45 degree or 63 degree, right-hand twill. Named for a Hebrew cloak or mantle popular during the Middle Ages. Made from most major fibres, alone or in blends, gabardine is a piece-dyed fabric much used in men and women's outer apparel. Some fabric may be skein or yarn dyed, or stock dyed. Weight runs from eight to fourteen ounces per yard and cotton yarn is used as the warp in the lower quality goods.

■ **garnett machine**

a type of carding machine, equipped with rollers and cylinders covered with metallic teeth, which is used to open up hard and soft waste textile products with a view to recycling them.

■ **garnetted yarn**

a yarn that has little bits ('garnets') of other fibres carded in. Usually the garnets are of a different colour, but they can also be from a different fibre.

■ **garnetting**
a technique for opening up hard and soft waste textile products with a view to recycling them.

■ **gassed yarns**
spun cellulose yarns passed over a heat source (or through a flame) to remove unwanted fibres on the surface. This gives a smoother surface but is not recommended at home. (Cellulose fibres are quite flammable.).

■ **gatt**
General Agreement on Tariffs and Trade, a multinational trade organisation established in 1947 and based in Geneva, Switzerland. GATT was superseded by the World Trade Organisation (WTO) in 1995.

■ **gauge**
a measurement most commonly associated with knitting equipment. It can mean the number of needles per inch in a knitting machine. However, in full fashioned hosiery and sweater machines, the number of needles per $1^1/_2$ inches represents the gauge.

■ **gel blocking**
a phenomenon- that occurs when the swelling of a super absorbent polymer blocks the passage of fluid into the centre of a fabric, thereby reducing the absorption capacity.

■ **Generalised System of Preferences (GSP)**
a system of tariff preferences operated by developed countries. The EU's scheme, introduced in 1971, was designed to foster the development of developing countries by granting them easier access to the EU market. Beneficiary countries granted GSP treatment were not required to contribute anything in return.

■ **geogrid**
a form of geotextile, which is a relatively stiff, mat-like material with open spaces in a rib structure.

■ **geomembrane**
an impermeable sheet of polymer, used in contact with soil or rock as part of a civil engineering operation. Geomembranes are used for such applications as lining reservoirs and waste dumps.

■ **georgette**
a sheer lightweight fabric, often made of silk or from such manufactured fibres as polyester, with a crepe surface. End-uses include dresses and blouses.

■ **geotextiles**
manufactured fibre materials made into a variety of fabric constructions and used in a variety civil engineering applications.

■ **gilet**
a waist- or hip-length garment, usually sleeveless, is fastening up the front. Sometimes made from a quilted fabric and designed to be worn over a blouse or shirt.

gilling

a commercial process called 'pin drafting' used to produce top fibres.

gingham

this fabric has dyed yarns introduced at given intervals in both warp and filling to achieve block or check effects. The warp and filling may often be the same, even-sided and balanced. Colour schemes range from conservative to gaudy, wild effects. Textures are around 64 x 56. Made from cotton, wool, worsted, nylon, etc.

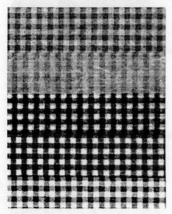

ginning

the mechanical process that removes the cotton fibres from the seed.

glass fibre

an inorganic fibre, which is very strong, but has poor flexibility and poor abrasion resistance. Glass will not burn and will not conduct electricity. It is impervious to insects, mildew and sunlight. Today, the primary use of glass fibre is in such industrial applications as insulation or reinforcement of composite structures.

glauber's salt

sodium sulphide. An acid used in dyeing to help the protein fibres to take colours evenly (levelling). Used in acid dyeing.

GMT

Glass Mat Thermoplastic offers better mechanical properties than injection-moulded reinforced thermoplastics, thanks to the higher residual length of the glass strands. GMT is widely used in automotive applications such as underbody shields, seat structures and front ends. GMT is obtained by consolidating a glass strand mat with a sheet of polypropylene. Chopping assembled roving and then needling the strands together obtain the glass mat. The intermediate product is delivered in the form of rigid sheets and subsequently moulded by the end user.

godet

a driven roller on a textile machine around which a yarn is passed in order to regulate its speed during the extrusion and further processing of certain man-made fibres. The roller may be heated in order to heat the yarn, which passes around it.

■ **GPD (gm/denier)**
a unit of force divided by the weight per unit length of a fibre, yarn or rope.

■ **grab tensile strength**
the strength at a specific width of fabric together with the additional strength contributed by adjacent areas.

■ **grade**
a measurement used in knitted garments that reflects the size of the needles used to knit the garment. The larger the gauge, the smaller the needle the finer the knit.

■ **grading**
classification of the unopened or untied fleeces according to fineness, staple length, character, soundness, etc.

■ **grandrelle**
a two-ply yarn composed of single yarns of different colours or contrasting lustre.

■ **granulation**
the process of forming new tissues.

■ **grey goods**
cloths, irrespective of colour, that have been woven in a loom, but have received no dry or wet-finishing operations. Grey goods are taken to the perch for the chalk marking of all defects, no matter how small. These blemishes must be remedied in finishing of the cloth. Material is converted from the grey goods condition to the finished state.

■ **grey wool**
fleeces with a few dark fibres, a rather common occurrence in the medium wools produced by down or black-faced breeds.

■ **grease wool**
wool in its natural condition as it comes from the sheep, either shorn or pulled. It contains a mixture of 'suint' and wool fats.

■ **great wheel or walking wheel**
turned by hand, very large (e.g., 50 inches in diameter), used for long draw on things like cotton (high twist). Instead of a flyer and bobbins, this wheel is a wheel-driven spindle.

■ **greige**
a term used to describe textile products prior to bleaching, dying or finishing. Some greige textiles may, however, contain dyed or finished yarns.

■ **greige goods**
an unfinished fabric, just removed from a knitting machine or a loom. Also called grey goods.

grist

the yards (or metres) per pound (YPP). So if you had a finished yarn that came up 890 YPP, one pound of yarn would equal 890 yards. The grist (or 'count') may range from 300 yds/lb to 3,000,000 yd/lb for a single filament of silk (theoretically).

grosgrain

a heavy, rather prominent ribbed fabric made from plain or rib weaves according to various combinations. The ribs will vary from a small number per inch to as high as 30 or 40 ribs to the inch. Made with silk or rayon warp and cotton filling, the fabric is rugged, durable and of the formal type, it is dressy and in place at formal gatherings. It finds much use in ribbons, vestments and ceremonial cloths.

GRP

Glass Reinforced Plastics.

guanaco

a protein fibre from the guanaco, a relative of the llama.

guard hair

the long, stiff, usually coarse fibre, which projects from the woolly undercoat of a mammal's pelt.

guipure

a lace construction produced by embroidering a thread pattern onto a fabric, the fabric being subsequently removed by chemical or other means to leave an open work lace.

gummy wool

scoured wool that still has some yolk in it.

gunny bag

a term of Sanskrit origin (Goni = sack) applied mainly to sacks and sacking made from jute but now used to describe other small bags made from other fibres, notably polypropylene.

habotai

a lightweight silk fabric commonly used for linings, hangings and underwear.

hackles

the comb for dressing flax or hemp.

hackling

cleaning the remaining woody particles and separation of the fibres. Drawing the flax fibres through the hackles does this. Traditionally, there are three sets of hackles that are used to progressively process the fibres. The previous step would be 'rippling'.

hair fibres

wool-like fibres from animals other than sheep, including the alpaca, llama, vicuna, cashmere goat, angora goat, angora rabbit and Bactrian camel.

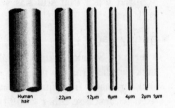

- half-blood wool

designation of a grade classification immediately below the fine grade.

- hand

the way the fabric feels when it is touched. Terms like softness, crispness, dryness and silkiness are all terms that describe the hand of the fabric.

- hand woven/hand loomed

fabrics, which are woven on either the hand or hand-and-foot power loom. They are admired because they express the individuality of the wearer.

- hand, handle

the reaction of the sense of touch, when fabrics are held in the hand. There are many factors which give 'character or individuality' to a material observed through handling. A correct judgement may thus be made concerning its capabilities in content, working properties, drape ability, feel, elasticity, fineness and softness, launder ability, etc.

- handkerchief

a square article made from any of the major textile fibres. It serves as a necessity or an adornment. It varies in size and may be decorated by the use of lace, a border, design or monogram. Often dyed or printed, the best grades are expensive and are usually made with a hand-rolled hem.

- handspun

yarns, which are, spun by hand using a spinning wheel or electric spinner.

- hand-washed wool

wool washed before it is shorn from the sheep.

- hank

a package of yarn from a reel, hopefully with the yardage and fibre content noted on a label. This may refer to a specified yardage, as in a hank of worsted yarn contains 560 yards, cotton and silk is 840 yards and linen is 300 yards.

hard twist
a yarn with increased twist.

harsh
a coarse, rough wool.

HDPE
High Density Polyethylene.

heat set finish
heat finishing treatment that will stabilise many manmade fibre fabrics so that there will not be any subsequent change in shape or size.

heather
a yarn that is spun using pre-dyed fibres. These fibres are blended together to give a particular look. (For example, black and white may be blended together to create a grey heathered yarn.) The term, heather, may also be used to describe the fabric made from feathered yarns.

heavy wool
wool that has considerable grease or dirt and will have a high shrinkage in scouring.

hemp
a coarse, durable baste fibre obtained from the inner bark of the hemp plant. Used primarily in twines and cordages and most recently apparel.

HEPA
High Efficiency Particulate Air (filtration).

herringbone
a variation on the twill weaves construction in which the twill is reversed or broken, at regular intervals, producing a zigzag effect.

herringbone twill
a broken twill weave giving a zigzag effect produced by alternating the direction of the twill. Same as the chevron weave. Structural design resembles backbone of herring. A true herringbone should have the same number of yarns in each direction, right and left and be evenly balanced. Thus, all herringbones are broken twills, but all broken twills are not true herringbones.

heterofilament
a filament made up of more than one polymer.

high pile
a pile in a fabric that is more than one eighth of an inch in height. When the pile is one eighth of an inch or less, the fabric is called a low-pile cloth.

HMPE
High Modulus Polyethylene.

hogget wool
hogget wool comes from sheep twelve to fourteen months old that have not been previously shorn.

The fibre is fine, soft resilient and mature and has tapered ends. Hogget wool is a very desirable grade of wool and because of its strength, is used primarily for the warp yarns of fabrics.

■ **hollow spindle system**

a system of yarn formation in which sliver or roving is drafted and the drafted twist less strand is wrapped with a yarn as it passes through a rotating hollow spindle. The binder or wrapping yarn is mounted on the hollow spindle and is unwound and wrapped around the core by rotation of the spindle. The technique may be used for producing a range of wrap spun yarns or fancy yarns.

■ **homespun**

in theory, this refers to rough, coarse, tweed-like fabric made with thick, uneven yarns and a plain weave. Obviously, not defined by a person who spends much time with good hand spinners.

■ **honeycomb**

a fabric structure in which the warp and weft threads form ridges and hollows, so as to give a cellular appearance.

■ **honeycomb waffle**

a raised effect is seen in this material which gives the effect of the cellular comb of the honeybee. The high point on the one side of the material is the low point on the reverse side. Care has to be used in manipulation. Used for draperies, jackets, skirts, women's and children's coats and dresses. Belongs in pique family of fabrics.

■ **hook**

the device used to pull the lead through the wheel's orifice.

■ **hopsack**

a modification of a plain weave in which two or more ends or picks weave as one.

■ **hopsacking**

popular woollen or worsted suiting fabric made form a 2-and-2 or a 3-and-3 basket weave. The weave effect is like that used for sacking to gather hops in the fields. Now made from other major fibres, hopsacking is used also for dress goods, jackets, skirts and blouses.

■ **horizontal lapping**

a process in which layers of web are laid horizontally, one on top of another, to form a multi layer structure.

■ **hound's tooth**

a medium sized broken-check effect, often used in checks, clear-finished worsteds, woollen dress goods, etc. The weave used is a four-end twill based on a herringbone weave with four ends to the right, followed by four ends to the left. The colour is completely surrounded by white yarn and the check is a four-pointed star, this two-up and two-down basic construction fabric is a staple in the fabric trade.

houndstooth check

a variation on the twill weave construction in which a broken check effect is produced by a variation in the pattern of interlacing yarns, utilising at least two different coloured yarns.

HPPE

High Performance Polyethylene.

HT

High Tenacity.

huckaback

a weave used principally for towels and glass-cloths in which a rough surface effect is created on a plain ground texture by weaving short floats, whereby warp floats are on one side of the fabric and weft floats are on the other.

hue

the pure spectrum colours commonly referred to by the 'colour names' - red, orange, yellow, blue, green violet.

hungry fine

a term used to describe a fine wool caused by poor nourishment as opposed to careful breeding.

HVAC

Heating, Ventilation and Air Conditioning.

hydroentanglement

a process for bonding a non-woven fabric by using high-pressure water jets to intermingle the fibres.

hydrophilic

a term used to describe a substance which tends to mix with or to be wetted by water.

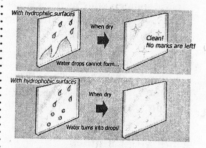

hydrophilic fibres

fibres that absorb water easily, take longer to dry and require more ironing.

hydrophilicity

the extent to which a substance is hydrophilic.

hydrophobic

a term used to describe a substance which tends to repel or not to be wetted by water.

identification test

any procedure for determining kinds of fibres, yarn construction, fabric construction or finish and colouring of textiles. Physical, chemical, microscopic and other methods may be used.

ikat

a traditional technique resulting in a streaky effect, created by tying and dyeing lengths of yarn before weaving.

impurity

any undesirable extraneous material present in a fleece or textile product.

inchworm

a pejorative term used to describe tense spinners who 'inch' their way through their fibres, often too close to the wheel orifice.

indigo

a blue dye from a variety of plants in the Indigofera family. Commonly used as a vat dye on both cellulose and protein fibres.

industrial textiles

a category of technical textiles used as part of an industrial process or incorporated into final products.

intarsia

a motif design knitted in solid colours into a weft knitted fabric.

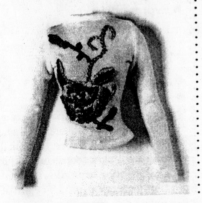

interfacing

fabrics used to support, reinforce and give shape to fashion fabrics in sewn products. Often placed between the lining and the outer fabric, it can be made from yarns or directly from fibres and may be either woven, non-woven or knitted. Some interfacings are designed to be fused (adhered with heat from an iron), while others are meant to be stitched to the fashion fabric.

interlining

an insulation, padding or stiffening fabric, either sewn to the wrong side of the lining or the inner side of the outer shell fabric. The interlining is used primarily to provide warmth in coats, jackets and outerwear.

interlock

the stitch variation of the rib stitch, which resembles two separate 1 x 1 ribbed fabrics that are interknitted. Plain (double knit) interlock stitch fabrics are thicker, heavier and more stable than single knit constructions.

interlock knit fabric

a special kind or type of eight-lock knit cloth, but it is generally described as a double 1x1 rib with crossed sinker wales. The fabric has a smooth surface on both sides possesses good wearing qualities and has less elasticity than ribs and does not develop prominent ribs when stretched in the horizontal direction. Fancy fabrics in this category are made with colour arrangements, needle set-out, tucking, missing and combinations of the foregoing. Used in sweaters and underwear.

■ **intermingled yarn**

a multifilament yarn in which cohesion is imparted to the filament bundle by entwining the filaments instead of or in addition to twisting. The effect is usually achieved by passing the yarn under light tension through the turbulent zone of an air-jet.

■ **intumescent system**

a flame retardant system which undergoes charring and foaming upon thermal degradation (for example, when exposed to an ignition source such as a flame). A blown protective cellular char is formed on the surface of the textile, providing protection from heat and flame.

■ **islands-in-the-sea**

a type of bi-component yarn in which one component polymer is formed, during extrusion, as longitudinal strands within the matrix of a second polymer.

■ **isotactic**

a type of polymer structure in which groups of atoms that are not part of the backbone structure are located either all above or all below the atoms in the backbone chain, when the latter are arranged all in one plane.

■ **jacob's fleece**

the natural brown shade of the Jacob's sheep.

■ **jacquard**

an intricate method of weaving invented by Joseph J.M. Jacquard in the years 1801-1804, in which a head-motion at the top of the loom holds and operates a set of punched cards, according to the motif desired. The perforations in the cards, in connection with the rods and cords, regulate the raising of the stationary warp thread mechanisms. Jacquard knitting is a development of the Jacquard loom and its principle. Jacquard fabrics, simple or elaborate in design, include brocade, brocatelle, damask, neckwear, eveningwear, formal attire, some shirting, tapestries, etc.

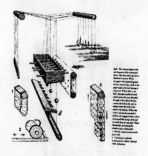

■ **jacquard knit**

a weft double knit fabric in which a Jacquard type of mechanism is used. This device individually controls needles or small groups of needles and allows very complex

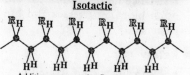

Addition occurs so that R groups are all on one side of the chain. Polymer has greatest crystallinity and mechanical strength.

and highly patterned knits to be created.

■ **jaspé**
a fabric characterised by a subtle striped effect.

■ **javanese**
a viscose cloth with a spun weft and filament warp, characterised by a dull sheen.

■ **jersey**
a plain stitch knitted cloth in contrast to rib-knitted fabric. Material may be made circular, flat or warp knitted, the latter type jersey is sometimes known as tricot. Used in dress goods, sportswear and underwear. Gives good service and launders very well. A very popular staple. Some fabric of this name is woven.

■ **jersey fabric**
the consistent interloping of yarns in the jersey stitch to produces a fabric with a smooth, flat face and a more textured, but uniform back. Jersey fabrics may be produced on either circular or flat weft knitting machines.

■ **jersey stitch**
a basic stitch used in weft knitting, in which each loop formed in the knit is identical. The jersey stitch is also called the plain, felt or stockinet stitch.

■ **jig dyeing**
this is done in a jig, kier, vat, beck or vessel in an open formation of the goods. The fabric goes from one roller to another through a deep dye-bath until the desired shade is achieved.

■ **joint venture**
a joint undertaking of a new, usually risky business in, for example, a developing country or in Eastern Europe.

■ **judo**
a structured cloth constructed in varieties of piqué weave and usually made in cotton.

■ **jute**
a baste fibre, chiefly from India, used primarily for gunny sacks, bags, cordage and binding threads in carpets and rugs.

■ **kaftan**
an oriental garment consisting of a long under-tunic tied at the waist by a girdle.

■ **kapok**
a short, lightweight, cotton-like, vegetable fibre found in the seedpods of the Bombocaceae tree. Because of its brittle quality, it is generally not spun. However, its buoyancy and moisture resistance makes it ideal for use in cushions, mattresses and life jackets.

■ **kelim**
turkish carpets with stylised geometric patterns.

Textile

kemp
a white, straight, opaque, coarse, non-felting, in-elastic fibre having a thick central medulla with hollow inter spaces. It will not take a dye, hence, its presence in wool is most objectionable. Often found around the head and legs.

keratin
a protein substance which is the chief component of wool fibre.

khaki
made of cotton, wool, worsted or linen, as well as from combinations of these fibres and the manmade fibres in blended materials. First used by the British armies as the official colour for uniforms at the time of the Crimean War 1853-1856. This ideal shade for field service finds only limited use in civilian dress-trouser, riding breeches, work clothes, children's play clothes, etc.

knit fabrics
fabrics made from only one set of yarns, all running in the same direction. Some knits have their yarns running along the length of the fabric, while others have their yarns running across the width of the fabric. Looping the yarns around each other holds knit fabrics together. Knitting creates ridges in the resulting fabric. Wales are the ridges that run lengthwise in the fabric, courses run crosswise.

knit-de-knit
a type of yarn texture in which a crimped yarn is made by knitting the yarn into a fabric and then heat-setting the fabric. The yarn is then unravelled from the fabric and used in this permanently crinkled form.

knitted geotextile
a geotextile produced by intermeshing loops from one or more yarns, fibres, filaments or other elements.

knop
a 'bunch' of fibres appearing along the length of yarn, giving a spot effect.

kpa (kilopascal)
the pressure produced by a force of 1,000 Newton applied, uniformly distributed, over an area of 1 m^2. (Used in textile testing as a measure of bursting pressure, 1 Kpa = 6.89 lbf/inch^2.).

KSI
Kilopounds per Square Inch (a unit of stress).

lace
the term comes from the old French, las, by way of Latin, laquens, which means a noose or to ensnare-rather well adapted to lace. A single yarn can produce a

plaited or braided fabric or article since it will interlace, entwine and twist in several directions to produce a porous material or lace and that the action is like that of several yarns entering the machine, this action is used in knitting as well.

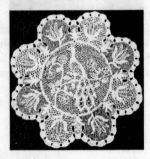

■ **lace effect**

refers to novelty fabrics of cotton, rayon, nylon or silk. Woven either in a leno pattern or in a heavy machine embroidery on a thin ground.

■ **lacquer**

a fabric finish which achieves a varnished look.

■ **ladder yarn**

a knitted tape yarn with the appearance of a ladder.

■ **lamb's wool**

wool shorn from lambs, usually when they are less than 7-8 months old. It is soft and has spinning qualities superior to fleeces of similar quality produced on older sheep.

■ **lame**

brocade, brocatelle or damask in which metallic (laminated) threads or yarns are interspersed throughout the fabric or one in which these threads have been used in the base construction. Most popular threads are those of copper, gold, silver, untarnished aluminium laminated with plastic is also used to provide rather brilliant colourings to simulate true metallic threads. It is a combination of butyrate acetate laminated to aluminium on both sides with the colour bonded into the adhesive.

■ **lanolin**

purified wool grease, chiefly a mixture of cholesterol esters. It is used in salves, cosmetic, grease paints and ointments.

■ **lawn**

made of carded or combed cotton yarn this light, thin cloth was first made in Laon, France. Comes in the white, solid colour or in prints. Satin stripes are often used for effect in the plain weave goods. Has a crisp and crease-resistant finish and is usually pre-shrunk prior to manipulating. It is sometimes crinkled to simulate plisse fabric. Crisper than voile but not as crisp as organdie in this family of cotton fabrics.

■ **lazy kate**

the device used to support full bobbins while plying. A fairly traditional design, involves two vertical posts that support the bobbins. Rather like a freestanding ladder. Another form, favoured by Schacht, has the bobbins supported horizontally with the addition of a breaking cord to control the backspin. Now, Alden Amos favours a vertical support with the addition of leather washers to help slow down the backspin. At this time, there is no image available.

■ **lea**

a form of measuring linen yarns in 300-yard increments and weighing one pound. A 4-lea skein would also weigh one pound but would be 1200 yards long.

■ **leaching**

the removal of a substance (such as a dyestuff) by a liquid which is in contact with the substance.

■ **leader**

this is the length of yarn attached to the centre core of a bobbin or shank of a spindle to aid in starting your yarn.

■ **leas ties**

also known as lees ties and lease ties. This is such an interesting term that pops in and out of textiles. The fact that a term used in measuring linen yarns is 'lea' makes one think leas ties came from that direction. So what are they? They are the short threads tied around hanks of yarn to help keep them from tangling while being washed, dyed and stored. They are also the short threads tied around a warp to allow you carry it from the warping board/mill to the loom. They serve the same process of keeping the threads in order. They are tied by running a thread at right angles to the warp/hank and interweaving through and coming back at opposite angles. Kind of a series of sideways figure 8s.

■ **leno weave**

a construction of woven fabrics in which the resulting fabric is very sheer, yet durable. In this weave, two or more warp yarns are twisted around each other as they are interlaced with the filling yarns, thus securing a firm hold on the filling yarn and preventing them from slipping out of position. Also called the gauze weave. Leno weave fabrics are frequently used for window treatments, because their structure gives good durability with almost no yarn slippage and permits the passage of light and air.

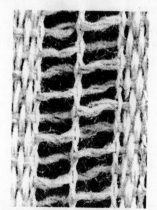

■ **leno woven fabric**

a fabric characterised by an open cellular appearance.

■ **leno-mesh**

a fabric in which warp threads have been made to cross one another between picks during leno weaving.

■ **level**

a dye term referring to even colour.

■ line flax
line fax is the long flax fibre that has been drawn off of the hackles. The finest preparation is often used for wet-spun linen, but line can also be dry spun.

■ line fleece
a fleece of wool midway between two grades in quality and length, which can be thrown into either grade.

■ linear density
the weight per unit length of a yarn or fibre. Units of linear density include decitex and denier.

■ linen
linen is the term used for fabric made from flax. Linen is generally

favoured for its fine, strong, cool-wearing properties. It is commonly cursed for its wrinkling. In knitwear, linen is combined with other natural or synthetic fibres for improved strength and resiliency.

■ lining
a fabric that is used to cover the inside of a garment to provide a finished look. Generally, the lining is made of a smooth lustrous fabric.

■ llama
llamas are a member of the camel family fibre originally from South America.

■ ILDPE
Linear Low-Density Polyethylene.

■ lock
a tuft or group of wool fibres that cling naturally together in the fleece.

■ loden
a thick heavy waterproof woollen cloth which is used to make garments, especially coats.

■ lofty wool
wool that is open, springy and bulky in comparison to its weight. This type of wool is desirable.

■ lOl
Limiting Oxygen Index. A measure of flammability. The level of oxygen in the oxygen/nitrogen atmosphere (expressed as a percentage) that must be present before a fibre will ignite and burn when exposed to flame.

■ long draw
this is a woollen-spinning technique.

■ long wool
wool from such breeds as the Lincoln, Leicester and Cotswold. It is large in diameter and up to 1215 inches in length.

■ loom
a machine used for weaving fabrics.

looper

an eyed stitch-forming element which carries an under thread or a cover thread on some types of sewing machine.

low wool

wool of low $1/4$ blood or lower in quality. Same as 'coarse wool'.

lowland wool

these breeds are characterised by producing wool that is generally coarser and only wavy or quite straight and of longer staple (over 4 inches). These wools are especially suited for the production of combed yarns, which are worked up into worsted fabrics.

IOY

Low Orientation Yarn.

lustre

the natural gloss or sheen characteristic of the fleeces of long-wool breeds. Fibres with a lot of lustre are often referred to as 'bright'.

lyocell

the generic name given to a new family of cellulose fibres and yarns that have been produced by solvent spinning. The process is widely regarded as being environmentally friendly and the product offers a number of advantages over traditional cellulose fibres.

lyocell fibre

a manufactured fibre composed of regenerated cellulose. Lyocell has a similar hand and drape as rayon, but is stronger, more durable and in many cases machine washable. It has a subtle lustre and is rich in colour. Lyocell possesses low shrinkage characteristics, as well as good absorbency and wrinkle resistant qualities.

macramé

knotted thread work.

madder

the roots of Rubia tinctorum used n vegetative (natural) dyeing to get a red or kind of a reddish brown colour.

madras

one of the oldest staples in the cotton trade, it is made on plain-weave background which is usually white, stripes, cords, or minute checks may be used to form the pattern. Fancy effects are often of satin or basket weave or small twill repeat. White filling is used. Yarn counts range from 40s to 60s in warp and filling while textures approximate 110 warp ends and 88 picks.

madras check

a colour-woven cotton fabric designed in colourful checks and usually associated with typical cotton checks from Madras (now Chennai) in India.

maiden

the name for the posts that support the flyer on a spinning wheel. Maiden and the base that supports them is called the mother-of-all.

mako cotton

very fine cotton spun from extra long staple Egyptian fibre.

maltinté

a yarn that is dyed unevenly to achieve an artificial aged effect.

man-made fibre

a fibre which is manufactured rather than occurring naturally. Man-made fibres can be further divided into cellulose or artificial fibres, which are made from naturally occurring polymers such as wood pulp and synthetic fibres, which are made from chemically derived polymers.

man-made filaments

filaments which are manufactured and which do not occur in nature.

maquiladoras

plants, common in Mexico and other Latin American countries, which process and assemble components or part-assembled goods made in the USA or another country and return the finished products to the USA or elsewhere for final sale. Usually, maquiladoras are in-bond assembly plants, which means that incoming goods can be freely imported without being liable to customs duty.

maquilas

see **maquiladoras**.

market class

the grouping of animals according to the use to which they will be put, such as slaughter or feeder.

marl yarn

a yarn consisting of two or more single ends of different colours plied together.

marocain

a crepe fabric with a weft-ways rib.

mass colouration

a method of colouring man-made fibres by incorporating a dye or colour in the spinning solution or melt before extrusion into filaments. Also known as dope-dyeing.

matelasse

a rather soft, double cloth or compound fabric which has a quilted surface effect. Made on Jacquard Looms, the heavier constructions are used for coverlets, draperies and upholstery. Lighter weight fabric finds use in dress goods, eveningwear and trimming. Matelassé gives effects such as blistered, puckered, quilted or wadded depending on the cloth construction used.

■ **matelassé crepe**

a soft, double or compound fabric with a quilted appearance which looks like two separate fabrics held together with creped threads on both sides. Named from the French verb meaning to pad or stud. Used for suits, coats, wraps, trimmings and dresses.

■ **mawata**

silk cocoons that have been simmered and opened onto a wooden frame.

■ **mcd/m²**

millicandela per m². The candela is the SI (Systéme International) unit for luminous intensity.

■ **mean fibre diameter**

the average diameter (thickness) of a group of fibres from an animal.

■ **mechanical bonding**

part of a production route for making non-woven, the web is cohered by using inter-fibre friction caused by physical entanglement. The entanglement can be caused by needles, high pressure water jets (hydro entanglement) or air jets.

■ **medium wools**

usually $1/4$, $3/8$ and $1/2$ blood wools, or wools grading 50's to 62's, or wools with a 24 to 31 micron count.

■ **meduiiated fibres**

fibres having more medulla (centre cell area), such fibres are coarse and uneven in diameter, harsh, low in elasticity.

■ **medulla**

the hollow, rounded cells that are found along the centre of the main axis of a fibre. They may run continuously along the length of the fibre.

■ **'MEG**

Monoethylene Glycol, a chemical intermediate used in the manufacture of polyester.

■ **melange**

a yarn produced from coloured printed tops or slivers. It is indistinguishable from a mixture yarn in that each fibre carries more than one colour.

■ **melt flow index**

an indication of the viscosity of molten polymer. The index serves to indicate the flow characteristics of a melt under given temperature and pressure conditions.

■ **melt spinning**

the conversion of molten polymer into filaments by extrusion through a spinneret and subsequent cooling of the extrude.

■ **meltblown**

part of a production route for making non-woven, extruded synthetic filaments are sucked by high pressure air jets from the die to form random length, very fine fibres which are deposited on to a belt.

melton

a heavyweight, dense, compacted and tightly woven wool or wool blend fabric used mainly for coats.

melt-spinning

a process in which the fibre-forming substance is melted and extruded into a gas or liquid, where it cools and solidifies. To form a non-woven, many fibres are created simultaneously and laid down as a web.

mercerising

a finishing process used extensively on cotton yarn and cloth consisting essentially of impregnating the material with a cold, strong, sodium hydroxide (caustic soda) solution. The treatment increases the strength and affinity for dyes and if done under tension, the lustre is greatly increased. This latter phase in now considered to be the heart of the process although not a part of John Mercer's original patent, discovered by accident in 1844. Mercerisation is done in skein form, on the warp, or in the piece, either entirely or in printed effects. Best results are noted in combed yarns.

merino wool

wool from the Merino sheep, with a mean fibre diameter generally of 24 microns or less.

mesh

a type of fabric characterised by its net-like open appearance and the spaces between the yarns. Mesh is available in a variety of constructions including woven, knits, laces or crocheted fabrics.

metallic

a manufactured fibre composed of metal, plastic-coated metal, metal-coated plastic or a core completely covered by metal. Metallic yarns are now made with Mylar polyester film as well as with acetate film. Mylar metallic yarns withstand higher temperatures and more rugged finishing and laundering than the acetate-type yarns.

metallic fibre

an inorganic fibre made from minerals and metals, blended and extruded to form fibres. The fibre is formed from a flat ribbon of metal, coated with a protective layer of plastic, which reduces tarnishing. Metal used in apparel fabric is purely decorative.

metallo-plastic

a yarn made from a synthetic or plastic material with a metallic appearance.

MFA

Multi Fibre Arrangement-a special protocol agreed by members of GATT as derogation from normal GATT rules. The MFA, which ran from 1974 to 1994, permitted members to establish quotas restricting textile and clothing trade which applied to specific supplying countries. GATT rules insist that all parties are to be treated equally. On January 1, 1995, the MFA was superseded by the Agreement on Textiles and Clothing (ATC).

■ **MFN**

Most Favoured Nation A basic principle of GATT which requires countries to treat imports from one GATT member no less favourably than imports from another GATT member.

■ **micro fibres**

the name given to ultra-fine manufactured fibres and the name given to the technology of developing these fibres. Fibres made using micro fibre technology, produce fibres which weigh less than 1.0 denier. The fabrics made from these extra-fine fibres provide a superior hand, a gentle drape and incredible softness. Comparatively, micro fibres are two times finer than silk, three times finer than cotton, eight times finer than wool and one hundred times finer than a human hair. Currently, there are four types of micro fibres being produced. These include acrylic micro fibres, nylon micro fibres, polyester micro fibres and rayon micro fibres.

■ **microfilament**

a continuous filament with a linear density approximately below 1 decitex. Some commercial filaments as coarse as 1.3 decitex are classified as microfilaments by their producers.

■ **micron**

a micron is $1/_{25,400}$ of an inch. The most accurate way of determining wool grades.

■ **micronaire value**

a measurement of cotton fibre quality. The micronaire value is a function of fibre fineness and maturity low values indicate fine and/or immature fibres, whereas high values indicate coarse and/or mature fibres. The micronaire value is determined in practice by measuring the resistance to airflow of a specified mass of fibres (in the form of a 'plug') confined in a chamber of a specified volume.

■ **microyarn**

a yarn consisting of several microfilaments.

■ **milling**

the operation of shrinking and felting a woollen fabric to make it thicker and denser. Also call 'fulling'.

■ **mineral colours**

actually they are not true dyes but are precipitated oxides or insoluble salts of chromium, iron, lead or manganese. Dull in appearance these colours are much used to colour awnings and comparable fabrics.

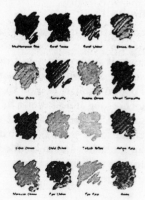

■ **miner's head**
an accelerating head used on walking (or great) wheels.

■ **mock leno**
a woven structure that imitates the appearance of leno weaves, i.e. it has an open structure.

■ **modacrylic fibre**
a manufactured fibre similar to acrylic in characteristics and end-uses. Modacrylics have a higher resistance to chemicals and combustion than acrylic, but also have a lower safe ironing temperature and a higher specific gravity than acrylic.

■ **modal**
a type of cellulose fibre having improved strength and modulus when wet.

■ **modulus**
a measure of the ability of a fibre to resist extension. Normally measured as the ratio of the stress (or load) applied on a yarn or filament to the elongation (strain) resulting from the application of that stress.

■ **mohair**
comes from the Angora goat, one of the oldest animals known to man. It is two-and-a-half times as strong as wool and outwears it. Foreign mohair is nine to twelve inches long and allowed a full year's growth before shearing. Uses include fancy goods, felt hats, linings, plushes and in blended yarns for use in men's and women's suiting fabrics.

■ **moiré**
a rippled effect created by applying heat and heavy pressure by means of rollers on a ribbed or corded fabric.

■ **moisture management (in textiles and garments)**
the process by which moisture is moved away from the skin and dispersed through a fabric to its outer surface. From here, moisture can evaporate, leaving both the skin and garment dry.

■ **moisture regain**
the amount of water a completely dry fibre will absorb from the air at a standard condition of 70 degrees F and a relative humidity of 65%. Expressed as a % of the dry fibre weight.

■ **moisture transport**
the movement of water from one side of a fabric to the other, caused by capillary action, wicking, chemical or electrostatic action.

■ **moisture vapour transmission**
the passage of water vapour, usually perspiration, through a fabric or membrane.

■ **moity wool**
wool that contains straw or other, non-seed-or-burr vegetable matter.

■ **moleskin**
1. heavy satin weave fabric made on a 5-end or an 8-end satin construction with the use of heavy, soft-spun filling in order to provide for a good napped surface

effect. Supposed to simulate the fur of a mole. Carded cotton yarn is used and the fabric is napped and sheared to provide what is actually a suede-effect.

2. a type of cotton goods 'fleece-lined' and having a soft, thick nap. Used as underwear in cold climates and in lining for the so-called sheep-skin-lined coats.

■ **monk's cloth**

a heavy weight cotton fabric utilising the basket weave variation of the plain weave. Used for draperies and slipcovers, monk's cloth is an example of 4 x 4 basket weave. It has poor dimensional stability and tends to snag.

■ **monoethylene glycol**

a chemical intermediate used in the manufacture of polyester.

■ **monofilament**

a single filament of a manufactured fibre, usually made in a denier

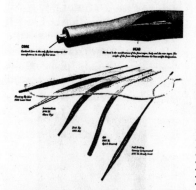

higher than 14. Monofilaments are usually spun singularly, rather than extruded as a group of filaments through a spinneret and spun into a yarn. End-uses include hosiery and sewing thread.

■ **monofilament yarn**

a yarn consisting of a single filament.

■ **mordant dyes**

a mordant is a substance used in dyeing to apply or fix colouring matter to a fibre, yarn or fabric, especially a metallic compound such as an oxide, which combines with the fibre and organic dye and forms an insoluble colour compound or lake in the fibre. Also known as Mordant-Acid Dyes or Chrome Dyes, they are closely related to Acid Dyes. Results are dull when compared with those from acid dyes. Exceptionally fast on wool and other animal fibres. Much used as well on carpeting, nylon and silk.

■ **mother-of-all**

the whole stand that supports the maidens, bobbin and flyer.

■ **mould (composites)**

a shaped former used to fabricate an article from a liquid or semi-solid under the effect of heat or pressure. Also used to describe the process of making the article in a mould.

- **mouliné**

    a type of two-colour twist yarn which gives a mottled effect in fabric.

- **mousseline**

    a general term for very fine, semi-opaque fabrics-finer than muslins-made of silk, wool or cotton.

- **mpa (megapascal)**

    the pressure produced by a force of 1 Newton applied, uniformly distributed, over an area of 1 mm^2.

- **MSW**

    Municipal Solid Waste.

- **mullen burst**

    the measured hydraulic bursting strength of a textile.

- **multifilament**

    a term applied to manmade yarns having many fine filaments. For example, 150 denier yarn with 40 filaments would be considered a standard filament count yarn in that denier size, but 150-denier yarn, with say 90 filaments, would be considered a multi-filament yarn.

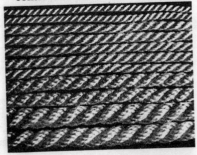

- **mungo**

    wool fibres recovered from old and new hard worsteds and woollens of firm structure. The fibres are less than 0.5 inch in length and owing to their reduced spinning and felting qualities, they are largely used in cheaper woollen blends. Mungo fibres are usually shorter than shoddy fibres.

- **mushy wool**

    wool that is lacking in character, dry and waste in manufacturing.

- **muslin**

    1. generic term for a wide variety of cotton fabrics, many of which are designated as muslin in connection with subordinate terms descriptive of the particular finish, weight, etc. in the sense that the word includes cloths ranging from lightweight sheers to the heavier weight, firmly woven goods such as sheeting.
    2. a white-goods finish on print cloth or sheeting which has been given a pure starched or backfilled finish to provide a dull, 'cloth' effect and hand. Muslin grey goods, for example, are finished in fabrics such as batiste, cambric, chintz, cretonne, lawn, longcloth, mercerised goods, plain muslin, nainsook, organdy, percaline, schreinered finish, etc.

nainsook

a lightweight plain weave cotton fabric, usually finished to create a lustre and a soft hand. Common end-uses are infants' wear, blouses and lingerie.

nanotechnology

research and technology development at the atomic, molecular or macromolecular levels (in the 1-100 nanometre range) aimed at creating and using materials which have novel properties and functions.

nap

a fuzzy, fur-like feel created when fibre ends extend from the basic fabric structure to the fabric surface. The fabric can be napped on either one or both sides.

natural dye

dye obtained from substances such as roots, bark, wood, berries, lichens, insects, shellfish and flowers.

natural fibre

fibre obtained from animal, vegetable or mineral sources, as opposed to those regenerated or synthesised from chemicals.

navajo ply

basically, a hand-crocheted loop used to create a thee-ply yarn.

needlepunching

a process for making a non-woven textile in which a continuous mat of randomly laid fibres or filaments is entangled with barbed needles. This causes matting and the production of a 'felt' textile.

nep

a small knot of entangled fibres commonly regarded as a fault but sometimes introduced as an effect.

net

an open fabric, which is created by connecting the intersections in a woven, knitted or crocheted construction to form a mesh-like appearance that won't ravel. End-uses include veils, curtains and fishnets.

netting

the knotting of threads into meshes that will not ravel. Chinese-type lace and fish net have a knot at every intersection. Knitted fabric may ravel or disentangle and the yarn may be used over again to make another fabric. Netting is done by hand or by machine.

nettle fibres

the nettle class of fibres comprise of Common Nettle (Urtica dioica), China Grass (Urtica nivea) and Ramie (or rhea).

neutral-premetalised acid dyes

much used on wool and other protein fibres, acrylics, modacrylics and nylon and ideal for use in blends. Much used in colouring apparel. Fair to excellent to light, good to excellent in washing and excellent in dry cleaning.

- **niddy-noddy**

a traditional, low-tech way to wind a skein and measure yarn. This image of the elusive niddy-noddy shows a very good quality, homemade one. For the less crafty or to dry damp skeins, consider making one out of $3/4$ or 1 inch pvc.

- **ninon**

a lightweight, plain weave, made of silk or manufactured fibres, with an open mesh-like appearance. Since the fabric is made with high twist filament yarns, it has a crisp hand. End uses include eveningwear and curtains.

- **nip**

a line or area of contact or proximity between two contiguous surfaces which move so as to compress and/or control the velocity of textile material passed between them.

- **noble comb**

used commercially in producing worsted yarn.

- **noil**

1. the short fibres taken from any machine operation in the processing of textile fibres. They are obtained mostly in carding and combing operations. The stock may be high in quality but very short in length, too short to admit its being manipulated into yarn by itself. Noil is worked in with longer staple fibres to make yarn. Some noil may be of medium or inferior quality.

2. short fibres that may be mixed in with longer staple woollen or worsted fibres in yarn manufacture, obtained from various frames.

- **noil silk**

waste fibres resulting from the processing of spun-silk yarn. F.T.C. rules require labelling as silk noil, silk waste or waste silk.

- **non-crushable linen**

plain weave cloth with highly twisted filling yarn or finished with resin to enhance elasticity. Has about the same uses as dress linen. Serviceable, durable, does not wrinkle, launders well.

- **non-woven**

a manufactured sheet, web or batt of directionally or randomly orientated fibres, bonded by friction and/or cohesion and/or adhesion, excluding paper and products which are woven, knitted, tufted, stitch bonded incorporating binding yarns or filaments or felted by wet-milling, whether or not additionally needled.

non-woven fabric

a textile structure produced by bonding and/or interlocking of fibres, accomplished by mechanical, chemical, thermal or solvent means and combinations thereof. The term does not include paper or fabrics that are woven, knitted, tufted or those made by wool or other felting processes.

non-woven geotextile

a geotextile in the form of a manufactured sheet, web or batt of directionally or randomly oriented fibres, filaments or other elements, mechanically and/or thermally and/or chemically bonded.

nostepenne

this deceptively simple-looking item is used to create a centre-pull ball.

novelty yarn

a yarn that is intentionally produced to have a special or unique effect. These effects can be produced by twisting together uneven single yarns, by using yarns that contain irregularities, or by twisting yarns that contain a colour variance. A slub yarn is an example of a novelty yarn.

numerical count system

a wool grading system. It divides all wools into 14 grades and each grade is designated by a number.

nylon

produced in 1938, the first completely synthetic fibre developed. Known for its high strength and excellent resilience, nylon has superior abrasion resistance and high flexibility.

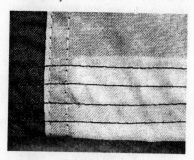

nytril

a manufactured fibre, most often used in sweaters or pile fabrics, where little or no pressing is recommended, as the fibre has a low softening or melting point. However, it has also been successfully used in blends with wool for the purpose of minimising shrinkage and improving the shape retention in garments.

occlusive

a covering (e.g. wound dressing) to protect and/or heal a moist wound.

OEM

Original Equipment Manufacturer. In the automotive supply chain, OEM's are vehicle manufacturers.

off-sorts

the by-products of sorting—shorts, britch wool, kemp, grey wool, stained wool, etc.

oilcloth

fabric that is treated with linseed-oil varnish to give a patent-leather effect. When used for table covers or shelf covering, it may be given a satiny sheen and finish. It comes in plain colours or printed designs. Other uses include waterproof garments for outerwear, book bags, covers, belts, bibs, pencil cases and other containers, surgical supplies, bags and luggage.

■ **olefin**

also known as polyolefin and polypropylene. A manufactured fibre characterised by its light weight, high strength and abrasion resistance. Olefin is also good at transporting moisture, creating a wicking action. End-uses include active wear apparel, rope, indoor-outdoor carpets, lawn furniture and upholstery.

■ **oleophilic**

a propensity to absorb oil.

■ **ombre**

a gradated or shaded effect of colour used in a striped motif. Usually ranges from light to dark tones of the one colour, such as from light blue to dark blue, can give usually, three or four different casts or tones of the one colour in the effect. Used in upholstery, some dress goods and decorative fabrics.

■ **ondé**

a fabric with a waved effect produced by calendaring or weaving.

■ **open wool**

wool that is not dense on the sheep and shows a distinct part down the ridge or middle of the back. Usually found in the coarser wool breeds.

■ **open-end spinning**

a spinning system in which sliver feedstock is highly drafted and thus creates an open end or break in the fibre flow. The fibres are subsequently assembled on the end of a rotating yarn and twisted in. Techniques for collecting and twisting the fibres into a yarn include rotor spinning and friction spinning.

■ **opening**

the second step in commercial wool processing (after sorting). The purpose is to open up the fleece in order that scouring will be more effective. Done with 'dusting'.

■ **organdy**

a stiffened, sheer, lightweight plain weave fabric, with a medium to high yarn count. End-uses include blouses, dresses and curtains/draperies.

■ **organza**

a very thin, but stiff plain woven silk, nylon, acrylic or polyester fabric used in evening wear, wedding attire for women, neckwear and trimming.

■ **organzine**

a silk yarn used for weaving or knitting. The yarn comprises single threads which are twisted, folded two, three or four-fold and finally twisted in the direction opposite to that of the single yarn.

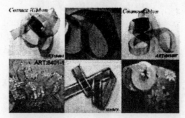

■ **orifice**

hole in end of flyer, directing the yarn to the bobbin, may also be a hook or pig's tail.

■ **osnaburg**

a tough medium to heavyweight coarsely woven plain weave fabric, usually made of a cotton or cotton/poly blend. Lower grades of the unfinished fabric are used for such industrial purposes as bags, sacks, pipe coverings. Higher grades of finished osnaburg can be found in mattress ticking, slipcovers, work wear and apparel.

■ **ottoman**

a tightly woven plain weave ribbed fabric with a hard slightly lustred surface. The ribbed effect is created by weaving a finer silk or manufactured warp yarn with a heavier filler yarn, usually made of cotton, wool or waste yarn. In the construction, the heavier filler yarn is completely covered by the warp yarn, thus creating the ribbed effect. End uses for this fabric include coats, suits, dresses, upholstery and draperies.

■ **outward processing**

a procedure whereby a company based in one country exports material to another country for additional processing and then re-imports the processed products for further treatment, for domestic distribution or for re-export. The most common form of outward processing involves the exporting of fabric from a high cost country to a low cost country for assembly or part-assembly into garments.

■ **overprinting**

colours or motifs printed over other colours already on the goods. Often done to alter shades or to tone-down certain vivid colours or effects that might be detrimental to the sale of the articles. Much used in floral, all over and splash prints.

■ **oxford**

soft, somewhat porous and rather stout cotton shirting given a silk-like lustre finish. Made on small repeat basket weaves, the fabric soils easily because of the soft, bulky filling used in the goods. The cloth comes in all white or may have stripes with small geometric designs between these stripes. Now is made from spun rayon, acetate and a few other manmade fibres. Oxford also means a woollen or worsted fabric which has a greyish cast made possible from a combination of black and white yarns or by the use of dyed grey yarn.

oxidation bases

one of these bases is aniline dye that is formed in the fibre by oxidation. Ideal for colouring fur, sheepskins and pile cloths, as well as in dyeing cotton for use in a wide range of fabrics. Finds much applications in print goods.

pa (pascal)

the pressure produced by a force of 1 Newton applied, uniformly distributed, over an area of 1 m². (Used in textile testing as a measure of bursting pressure.).

package dyeing

yarns are dyed while on cones, cakes, cheeses or in the conventional or standard layout or set-up.

padding (finishing)

the impregnation of a substrate with a liquor or paste followed by squeezing, usually by passing the substrate through a nip, to leave a specific quantity of liquor or paste on the substrate.

paisley

a teardrop shaped, fancy printed pattern, used in dresses, blouses and men's ties.

pan

polyacrylonitrile: a precursor for making carbon fibres.

panne

a satin-faced, velvet or silk material named from the French for 'plush', which has a high lustre made possible by the tremendous roller-pressure treatment given the material in finishing. Panne velvet is often referred to as panne and is a staple silk fabric.

papermakers felts

textile based felts used to extract water during the process of papermaking.

partially oriented yarn

a continuous synthetic filament made by extruding a polymer so that a substantial degree of molecular orientation is present in the resulting filaments, but further substantial molecular orientation is still possible. The resulting yarn will usually have to be drawn in a subsequent process in order to orient the molecular structure fully and optimise the yarn's tensile properties.

passementerie

an open-work braid technique, traditionally used for furnishing braid.

pattern

1. an outline of a garment on paper. It embodies usually all the pieces necessary to cut a complete garment from material.
2. a single repeat of a weave formation.

■ PBT

polybutylterephalate, a type of polyester used as an engineering plastic and, for specialist uses, in the form of a fibre.

■ PCB

Polychlorinated Biphenyl. PCBs are a group of toxic, chlorinated aromatic hydrocarbons used in a variety of commercial applications, including paints, inks, adhesives, electrical condensers, batteries and lubricants. PCBs are known to cause skin diseases and are suspected of causing birth defects and cancer.

■ peachskin

the term used to describe the soft surface of certain textiles which feels like and has the appearance of the skin of a peach.

■ peasant combs

single-pitch and 2-pitch hand combs used to produce a semi-worsted fibre preparation.

■ peau de soie

a heavy twill weave drape able satin fabric, made of silk or a manufactured fibre and used for bridal gowns and eveningwear.

■ pebble effect

fabric with a rough, granite-like, irregular or pebble effect on the face of the goods. Most of the cloth is some type of crepe fabric.

■ pelt

the skin from a slaughtered sheep before the wool on it has been pulled or processed into a sheepskin.

■ pencil locked

a fleece with narrow staples or lock formation indicates an open fleece that has less density and probably more vegetable matter. This type of lock formation is genetic and is passed on to offspring.

■ pepper and salt

a fabric with a speckled effect, often black and white.

■ percale

1. dress percale is a medium-weight, printed cotton cloth with a firm, smooth finish. Made from plain weave the texture is around 80-square. Used for women's and children's dresses, aprons and blouses. Used interchangeably with the word, calico.
2. sheet percale is fine, smooth, lustrous and highly textured with a count of 90 to 100 yarns each way for a total of 180 to 200 threads in the goods, a very high grade sheeting.
3. percaline is the name applied to a summer coat or suiting fabric made of cotton. Usually piece dyed, it is given a glazed or a Moire finish. Popular, at times, for boleros.

■ **permanent finish**

a comparatively much-used term given a number of materials for which some particular claim is made. Examples could include Moire or watermarked effect on faille, taffeta, organdy and dimity the smoothness on broadcloth, embossed fabrics, some crepe effects, glazed fabrics, etc. The term, as well, implies cloths which are crease-resistant, shrinkage-resistant and wear resistant. In reality, however, the safer term to use is 'Durable Finish'. Actually, very few finishes truly qualify as being permanent for the life of the fabric or garment.

■ **permanent press, durable press**

terms used to describe a garment that will retain its shape-retaining properties throughout its career. Features include sharp creases, flat seams, smooth surface texture and appearance on the goods and seams which are free from puckering.

■ **permeability**

the ability of a textile to allow air or water vapour to pass through it.

■ **permittivity**

the volumetric flow rate of water per unit of cross-sectional area, per unit head, under laminar flow conditions, in the normal direction through a geotextile.

■ **PET**

Polyethylene Terephthalate, the most common form of polyester.

■ **PHA**

polyhydroxyalkanoate.

■ **Phase-Change Materials (PCMs)**

materials which change their state of matter, usually from solid to a liquid or vice versa. PCMs are often used in textiles which are intended for sport and exercise to help maintain comfort and constant body temperature.

■ **PHB**

polyhydroxybutyrate.

■ **pheromone**

a chemical substance secreted externally by certain animals which affects the behaviour or physiology of other animals of the same species.

■ **photographic printing**

photographic prints can be transferred to fabric by the use of photoengraved rollers. Various ways are used to obtain the result, all adapted from colour-printing on paper. Red, yellow and blue, the primary colours are much used to obtain a host of colour-effects.

■ **PHV**

polyhydroxyvalerate.

pick | pigment colours

pick
in weaving, a weft yarn. In spinning, the process of opening out the fibre to help in the cleaning and processing. This process (picking) allows a lot of vegetable matter to drop out of fleece.

pick glass
a single or double lens glass used in analysing and dissecting cloth. Comes in 1/4 inch and 1/2 inch and 1 and 2 inches sizes. For good results, a one-inch glass should be used. The device is hinged so that it may be readily folded up. Also known as a counting-glass, linen-tester or pick counter.

picker
this is a rather terrifying tool used to help open out washed fleece prior to carding. Basically it is a crescent-shaped swing that rocks in a cradle. Between the base of the picker and the bottom of the swing, there are a series of nails that catch the fibre and open it out. This tool accomplishes much the same process as hand-picking your fibre, but much faster. As with any machinery, use with extreme caution.

picking
on a picker machine there is a continued cleansing and blending of the fibres being treated. The final result from the last picker frame in the set is in lap formation with the stock ready for carding. Cotton laps, for example, are about 45-inches in width and weigh about 45 pounds.

piece dyeing
the dyeing of fabric in the cut, bolt or piece form. It follows the weaving of the goods and provides a single colour for the material, such as a blue serge, a green organdy.

pieces
the skirting and other less-desirable pieces of wool removed from the fleece. (Australian term).

pigment
a very finely divided white or coloured solid which is present in or on a fibre to impart dullness or colour to the fibre. Titanium dioxide or barium sulphate are white pigments often used for dulling manmade fibres. In dope-dyeing or spun-dyeing coloured pigments are added to the polymer solution or melt before the yarn is spun.

pigment colours
insoluble in water and the colour has to be fixed onto the fibre by use of resinous-binders insolubilised by a curing treatment at high temperatures. Used mostly on cotton and acetate, rayon and some other manmade fibres. These dyes, generally speaking, colour by dyeing or printing, just about all types of fibres and blends. Light and medium shades

of sailcloth's and many types of dress goods use Pigment colours. Good to excellent in light-fastness but may be poor in crocking or rubbing.

■ **pile fabric**

one in which certain yarns project from a foundation texture and form a pole on the surface. Pile yarns may be cut or uncut in the fabric. Corduroy and velveteen are examples of cut filling pile fabrics, velvet is an example of a cut warp pile fabric while Turkish towelling or Terry cloth is an uncut pile material.

■ **pile knit**

a type of knit construction which utilises a special yarn or a sliver that is interloped into a standard knit base. This construction is used in the formation of imitation fur fabrics, in special liners for cold weather apparel such as jackets and coats and in some floor coverings. While any basic knit stitch may be used for the base of pile knits, the most common is the jersey stitch.

■ **pile weave**

a type of decorative weave in which a pile is formed by additional warp or filling yarns interlaced in such a way that loops are formed on the surface or face of the fabric. The loops may be left uncut or they may be cut to expose yarn ends and produce cut pile fabric.

■ **pill**

the entangling of fibres during washing, dry cleaning, testing or in wear to form balls or pills which stand proud of the surface of a fabric and which are of such density that light will not pass through them (so that they cast a shadow).

■ **pilling**

the tendency of some yarns to form little balls of short, tangled fibres on the surface. This tendency can be reduced (or removed) by removing the short fibres (combing) or by adding additional twist.

■ **pillowcase linen**

plain weave, high count, good texture, bleached. Yarn is very smooth and has high count of turns of twist per inch. Launders easily and well, sheds dirt, has cool feel and appearance, is strong and durable.

■ **pima**

a type of long-staple cotton.

■ **pina**

the vegetable leaf fibre from the pineapple plant.

■ pin-drafting

a system of drafting in which the fibres are oriented relative to one another in the sliver and are controlled by rolls of pins between the drafting rolls. It is primarily used for long fibres in the semi-worsted and worsted spinning systems.

■ pinwale

a very narrow ridge or rib in a fabric (from 16 to 23 wales to the inch). Example-pinwale corduroy.

■ pique

a medium-weight fabric, either knit or woven, with raised dobby designs including cords, wales, waffles or patterns. Woven versions have cords running lengthwise or in the warp direction. Knitted versions are double-knit fabric constructions, created on multi-feed circular knitting machines.

■ PLA

Polylactic Acid: a synthetic polymer formed from plant-based material and used as the starting material for a new range of fibres.

■ plaid

1. a pattern consisting of coloured bars or stripes which cross each other at right angles, comparable with a Scottish tartan. Plaid infers a multi-coloured motif of rather large pattern repeat, the word check refers to similar motifs on a smaller scale and with fewer colours.
2. a rectangular piece of fabric or a garment, having a plaid or tartan design, worn by both sexes in Scotland in lieu of a cloak.
3. a woollen cloth with a tartan motif in which plaids, two of them, are woven into the goods, also called over plaid.
4. a woven or printed pattern in a tartan with some appealing cross-barred effect. Plain weave is the most common of the fundamental weaves. Each filling yarn passes successively over and under each warp yarn, alternating each row. Sometimes called the one-up and one-down weave. Used for muslin, print cloth, taffeta, voile, sheeting, etc.

■ plain wool

wool lacking character with few crimps.

■ plant fibre

a fibre generated from a plant, e.g., cotton, flax.

■ plasma

ionised gas.

■ plastic

any material, natural or synthetic, which may be fabricated into a variety of shapes or forms, usually by the application of heat and pressure. A plastic is one of a large group of organic compounds synthesised from cellulose, hydrocarbons, proteins, or resins and capable of being cast, extruded or moulded into various shapes. From the Greek plastikos, which means 'fit for moulding'.

■ **plating**

a process for making a knitted fabric from two yarns of different properties-one on the face of the fabric, the other on the back.

■ **plied yarns**

yarns produced by two or more singles have been twisted together.

■ **plissé**

a lightweight, plain weave, fabric, made from cotton, rayon or acetate and characterised by a puckered striped effect, usually in the warp direction. The crinkled effect is created through the application of a caustic soda solution, which shrinks the fabric in the areas of the fabric where it is applied. Plissé is similar in appearance to seersucker. End-uses include dresses, shirting, pyjamas and bedspreads.

■ **ply**

a single unit of yarn. A 2-ply yarn would involve taking two singles and then plying them in the opposite direction they were originally spun.

■ **plying**

the process of taking multiple singles and twisting them back against themselves.

■ **pole weave**

this weave requires two or more warps and one filling or two or more fillings and one warp. The extra warp or filling is called the 'pile' warp or filling and forms the loops on the face of the goods. If the loops are cut, it is done by knife blades which are attached to the loom, except in fine velvets, in which there is a separate process. Pile fabric should not be confused with napped fabric.

■ **polishing**

usually done on commercial sewing thread, the process involves burnishing the plied yarn.

■ **polyester**

a manufactured fibre introduced in the early 1950s and is second only to cotton in worldwide use. Polyester has high strength (although somewhat lower than nylon), excellent resiliency and high abrasion resistance. Low absorbency allows the fibre to dry quickly.

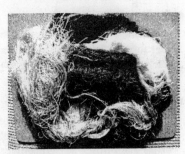

■ **polypropylene**

also known as polyolefin and olefin, a manufactured fibre characterised by its light weight, high strength and abrasion resistance. Polypropylene is also good at transporting moisture, creating a wicking action. End-uses include active wear apparel, rope, indoor-outdoor carpets, lawn furniture and upholstery.

■ **pongee**

the most common form is a naturally coloured lightweight, plain weave, silk-like fabric with a slub effect. End-uses include blouses, dresses, etc.

■ **poplin**

a fabric made using a rib variation of the plain weave. The construction is characterised by having a slight ridge effect in one direction, usually the filling. Poplin used to be associated with casual clothing, but as the 'world of work' has become more relaxed, this fabric has developed into a staple of men's wardrobes, being used frequently in casual trousers.

■ **post-cure**

a type of durable press finish in which the finish is applied to the fabric by the mill but the garment manufacturer completes the cure of the finish by applying heat, using an oven or press or both to the completed garment.

■ **PPS**

Polyphenylene Sulphide.

■ **pre-cure**

a type of durable press finish in which the finish is applied to the fabric and set or cured, through the use of heat by the mill prior to shipment of the fabric to be made into garments.

■ **precursor**

raw materials used in a controlled pyrolysis process to make carbon fibres.

■ **prepreg**

an assembly of fibres impregnated with resin, prepared for performing into a composite shape before the curing process used to set the resin.

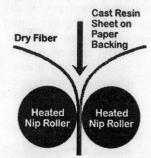

■ **pre-shrunk**

fabrics or garments that have received a preshrinking treatment. Often done on cottons to remove the tendency for cloth to shrink when washed or laundered. Worsted and woollens are also shrunk before cutting the fabric for use in a garment to prevent further shrinkage. The per cent of residual shrinkage must be indicated on the label of the goods or garments thus treated.

■ **press**

1. any machine that presses or smoothes fabrics or garments.
2. a machine of which there are several types, used to press or compress raw materials.
3. to iron in the home or commercial laundry.
4. to squeeze liquid out of cloth by press rollers.
5. the opposite of 'to pull'.

■ **press cloth**

fibre cloth made of camel's hair, cotton and wool, depending on use. Hair fibre is found in varying percentages in all press cloths.

■ **pressure dyeing**

one of the popular methods of colouring textiles whereby the material and the dye liquor are held under steam pressure in a closed jig, kier or vessel. The plan is to provide a quicker dyeing when temperatures necessary are above the boiling point of 212 °F.

■ **primaries**

the three basic 'hues'. The Painters Primaries include red, blue and yellow. This is the palate one learns in school. The Printers Primaries include magenta, cyan and yellow. This is the palate used for colour printing. The Light Primaries include red, blue and green. This is the palate used on your computer screen.

■ **primary colours**

red, yellow and blue, from which pigments of these colours may be mixed to make many other colours.

■ **primitive**

a fleece with both long and short fibres.

■ **prince of Wales**

a large-scale check, typified by a reversing effect ground with an over check.

■ **print cloth**

a medium weight, plain weave cloth made of carded yarns that range from 28s to 42s in count or size. There is a host of constructions and qualities of the fabric whose pick counts range from 64 x 64 or 64 square to 80 square. Widths vary from 38 1/2 inches to 40 inches. Print cloths are converted into more different finishes than any other grey goods fabric. Millions of yards are printed annually while other millions receive the white goods finish as noted in cambric, lawn, long cloth, muslin, nainsook, organdy, etc. Large amounts of the goods are used in the grey condition for bags and containers and as the base fabric for coated materials.

■ **printing**

producing patterns, designs or motifs of one or more colours onto fabric. Several methods and techniques are used in printing and these follow.

- **production sequence**
  shearing, sorting, opening, cleaning, carding, drawing, possibly combing, possibly roving, twisting or spinning.

- **progressive bundle system**
  a system traditionally employed in apparel production where the task of assembling the garment is broken down into small operations and bundles of work are progressed down the production line through each operation in sequence until the assembly process is complete.

- **progressive shrinkage**
  shrinkage which results from repeated washing, laundering or dry cleaning, shrinks more after each successive treatment.

- **protection (geotextiles)**
  the use of a geotextile as a localised stress reduction layer to prevent or reduce damage to a given surface or layer. This refers mainly to the protection of geomembranes from damage due to sharp rock particles or other materials in landfill applications.

- **protein fibre**
  a fibre composed of protein, including such naturally occurring animal fibres as wool, silk, alpaca, llama and other hair and fur fibres.

- **provençal**
  small stylised floral typical of the Provence region of France.

- **PTA**
  Pure Terephthalic Acid, used in the manufacture of polyester.

- **PTTFE**
  Polytetrafluoroethylene.

- **PU**
  polyurethane.

- **pulled wool**
  wool pulled from skins of slaughtered sheep. The wool is pulled from the skins after treatment of the fleshy side of skins with a depilatory. Pulled wool should not be confused with 'dead wool'.

- **pultrusions**
  composites produced by drawing resin-coated filaments through a pressure die.

- **puncture resistance (geotextiles)**
  the extent to which a geotextile is able to withstand or resist the penetration of an object without perforation.

- **puni**
  a puni is a tighter-than-normal rolag traditionally used with cotton.

- **purebred**
  an animal of pure breeding, registered or eligible for registration in

the heard book of the breed to which it belongs.

- **purity**
refers to the absence of dark fibres, kemp or hair.

- **purl stitch**
a basic stitch used in weft knitting, which produces knit fabrics that have the same appearance on both sides. The purl stitch is frequently used in combination with the jersey and rib stitches to produce a knitted fabric design. Sweaters, knitted fabrics for infants and children's wear, knitted fabrics for specialised sportswear and bulky knit fabrics are commonly made using the purl stitch.

- **push-pull fabrics**
bi-component fabrics composed of a non-absorbent hydrophobic material, usually polyester, on the inside (worn next to the skin) and an absorbent hydrophilic material, usually nylon, on the outside.

- **PVA**
Polyvinyl Alcohol.

- **PVC**
Polyvinyl Chloride.

- **PVDF**
polyvinyldifluoride.

- **PVF**
polyvinylfluoride.

- **pyrolysis**
a process in which chemical compounds are decomposed at high temperatures.

- **qiviut**
down undercoat fibre from the musk ox.

- **quality**
refers to the degree of fineness.

- **quarter-blood wool**
one of the grades in the standards for wool.

- **quasi-isotropic laminates**
composite laminates with almost isotropic mechanical properties.

- **quilt**
1. usually a bed covering of two thickness of material with wool, cotton or down batting in between for warmth.
2. used for jackets and linings of coats.
3. also the sewing stitch used to make a quilt.

quilting

a fabric construction in which a layer of down or fibrefill is placed between two layers of fabric and then held in place by stitching or sealing in a regular, consistent, all-over pattern on the goods.

quota

a quantitative restraint imposed by an importing country on an exporting country or established by agreement between the trading partners which is designed to limit shipments of a product from the exporting to the importing country.

rabbit hair

this hair is used in combination with other fibres. It is soft and lustrous and in the better quality fabrics enough hair may be present to justify the use of the term. Much used in varying percentages in wool and blend fabrics.

raffia

a fibre obtained from the leaves of the raffia palm.

raglan

named for Lord Raglan, a British general, this popular style of coat is a loose-fitting garment which may or may not have sleeves. Called a cape when the article is sleeveless. A feature of the raglan model is that there is plenty of room for arm movement, highly desirable in driving a car.

ramie

the baste fibre produced from the Asian urticaceous shrub Boehmeria nivea or Boehmeria tenacissima. Used to also be called 'rhea' or 'China grass'. The fibre is white, soft, lustrous and slightly coarser than flax (linen) when degummed and bleached. Ramie fabrics are strong, smooth and durable.

random dyeing

colouring only certain designated portions of yarn. There are three ways of doing this type of colouring:
1. Skeins may be tightly tied in two or more places and dyed at one side of the tie with one colour and at the other side with another one.
2. Colour may be printed onto the skeins which are spread out on the blanket fabric of the printing machine.
3. Cones or packages of yarn on hollow spindles may be arranged to form channels through which the yarn, by means of an air-operated punch and the dyestuff are drawn through these holes by suction. The yarn in the immediate area of the punch absorbs the dye and the random effects are thereby attained.

■ **range wool**

wool produced under range conditions in the West or the Southwest. With the exception of Texas and California wools, it is usually classified as territory wool.

■ **raschel**

a two-needle warp knitting system.

■ **raschel knit**

a warp knitted fabric in which the resulting knit fabric resembles hand crocheted fabrics, lace fabrics and nettings. Rachel warp knits contain inlaid connecting yarns in addition to columns of knit stitches.

■ **ratine**

a cloth with a rough surface, which has been achieved by finishing and/or the use of fancy yarns.

■ **raw fibres, raw material**

material in natural condition made suitable for manipulation into a product. Examples include raw cotton, raw wool, silk 'in-the-gum', etc.

■ **raw silk**

continuous silk containing no twist that has been drawn off of cocoons. The fibres are often un-degummed.

■ **raw stock dyeing**

dyeing of fibre stock precedes spinning of the yarn. Dyeing follows the degreasing of the wool fibres and drying of the stock.

■ **raw wool**

wool in the grease, as shorn from the sheep. Same as 'grease wool'.

■ **rayon**

a manufactured fibre composed of regenerated cellulose, as well as additional manufactured fibres composed of regenerated cellulose in which the substitutes have replaced not more than 15 per cent of the hydrogen of the hydroxyl groups. Rayon Processes.
1. Cuprammonium-Filaments made of regenerated cellulose coagulated from a solution of cellulose in ammoniacal oxide.
2. Viscose-Filaments made of regenerated cellulose coagulated from a solution of cellulose xanthate.

■ **reactive dyes**

they actually bond-in the colourant. Provide bright colours on cottons and can dye acrylics, nylon, silk and wool and blends of these fibres. Also used for printing cotton fabrics. Rated good to very good to light and washing, but fugitive to chlorine-base bleaches.

- **reclaimed wool**

  wool that is reclaimed from new or old fabrics.

- **recovery**

  a basic property of stretch yarn and refers to the degree to which a yarn returns to its relaxed position after stretching. Rapid and complete recovery prevents bagging or sagging and is very important in many types of articles, especially in stretch pants.

- **redox**

  a type of chemical reaction in which one of the reagents is reduced, while another is oxidised.

- **redox agent**

  a substance which promotes redox reactions.

- **reed**

  a device consisting of several wires closely set which separate warp threads in a loom. The reed determines the spacing of the warp threads, guides the weft-carrying device and beats up the weft against the fell of the cloth.

- **reed width**

  the width of the fabric in the reed.

- **reeled silk**

  a long strand made of silk reeled from a number of cocoons and not twisted or spun.

- **regain**

  the ratio of the weight of water in a material to the oven-dry weight of the material.

- **regenerated fibre**

  these are fibres created by modified natural fibres. The cellulose regenerated fibres include rayon and acetate. The protein-regenerated vegetable fibres include soybean (soylon), peanut (ardil) and corn (vicara). The protein-regenerated animal fibres include casein (aralac), gelatin and albumin.

- **reinforcement (geotextiles)**

  the ability of a geotextile to reduce stresses or contain deformation in geotechnical structures. The geotextile enhances the shear strength of the soil mass by adhering to the adjacent soil layers. The geotextile layers are normally placed across the potential failure planes to carry the tensile forces, which cannot be borne by an unreinforced soil mass.

- **rejects**

  off-grades thrown to one side by the wool grader, fleeces with excessive black fibres, kemp, dead fibres, vegetable matter, etc.

- **relief printing**

  a method upon which the 'hill areas' of the engraved plate are inked for printing. It is actually the reverse of Intaglio Printing.

- **repeat**

  1. an entire completed pattern for design and texture. Repeats vary in size considerably, depending on

the weave, type of material, texture and the use of the cloth.
2. the form which indicates the size of the weave and the number of threads that the weave contains in both the warp and filling. A weave may be repeated any reasonable number of times on paper.

■ **repellency**

the ability of a fabric to resist such things as wetting and staining by water, stains, soil, etc.

■ **repellent treatments**

any one of a great number of treatments that may be applied to fabrics and garments to make them repellent to moths, water, mildew moisture, perspiration, etc. They may or may not be durable in nature. Some disappear after the first washing and laundering, others will give good service, at times almost to the point of permanency. A very few may be classed as permanent.

■ **reprocessed wool**

scraps and clips of woven and felted fabrics made of previously used wool. These remnants are garneted, that is, shredded back into a fibrous state and used in the manufacture of woollens.

■ **resilience**

the power of recovery to original shape and size after removal of the strain which caused the deformation. A fibre may possess this quality to spring back to its original state after being crushed or wrinkled. Resilience is sometimes referred to as memory.

■ **resin**

the name commonly applied to chemical compounds used to impart wash-and-wear and durable press properties to fabrics which contain cellulose fibres.

■ **resin bonding**

this method is the more popular since it can be applied directly to the web from an aqueous dispersion as soon as it leaves the web-forming equipment. The dispersion may be applied in spray form, foam form or by printing or padding onto the web. The web is then dried and heat cured and may be calendared. Drying and curing are done on cans or in curing ovens. Temperatures range from 200 to about 400 °F. Rebinding is often done to further strengthen the finished fabric.

■ **resist dyeing**

treating yarn or cloth so that in any subsequent dyeing operation the treated portions resist the dye and do not absorb it at all.

■ **resist printing**

fabric is 'dyed' with a tannin-mordant paste and then the desired areas to be used for the motif are stripped of the covering, leaving the areas white. The fabric is then piece-dyed and some or all of the white areas are coloured by this direct method of printing.

■ **resist treatment**

a treatment applied to part of a fabric which causes the area treated to resist the take-up of dye.

retention

the weight of fluid remaining after a freely swollen fibre, yarn or fabric is subjected to a pressure of 0.5 lb/in^2.

reticulated foam

reticulated foams differ from conventional foams in their cell structure. Reticulation is a process in which cell membranes are destroyed in a controlled explosion and then fused with the cell ribs. Reticulation produces open-celled foam which is especially suitable for the filtration and purification of air and liquids.

retting

this is the process in flax production that weakens the fibres in the flax plant. Several retting methods are used:
1. Dew or Grass Retting. Small bundles of the uprooted flax plants are left outdoors for 3-5 weeks.
2. Pond Retting. Small bundles are left submerged for 4-8 days. Many books refer to an unpleasant stench as a side effect of this process.
3. Stream Retting. Small bundles are anchored in a body of moving water. This is the quickest and the cleanest of the processes.
Apparently efforts are underway to perfect a process using enzymes to replicate the dew process. Retting occurs in flax fibre production after 'rippling' and before 'breaking'.

reused wool

these are cleaned and shredded into fibres again and then blended to make utility fabrics.

revetment

a support structure in civil engineering made of riprap (coarse armour stone) or concrete.

reworked wool

wool that has been previously used. Also called 'shoddy' and 'mungo'.

rib

usually a straight raised cord, formed in a weave by threads that are heavier than the others, lengthwise, crosswise or diagonal. Many knitted fabrics are ribbed lengthwise.

rib knit

a basic stitch used in weft knitting in which the knitting machines require two sets of needles operating at right angles to each other. Rib knits have a very high degree of elasticity in the crosswise direction. This knitted fabric is used for complete garments and for such specialised uses as sleeve bands, neckbands, sweater waistbands and special types of trims for use with other knit or woven fabrics. Lightweight sweaters in rib knits provide a close, body-hugging fit.

rib weave

one of the plain weave variations, which is formed by using.
1. heavy yarns in the warp or filling direction.
2. a substantially higher number of yarns per inch in one direction than in the other.
3. several yarns grouped together as one. Rib fabrics are all characterised by having a slight ridge effect in one direction, usu-

ally the filling. Such fabrics may have problems with yarn slippage, abrasion resistance and tear strength. Examples of this construction include broadcloth, poplin, taffeta, faille, shantung and cord fabric.

■ ribbed fabric

made with two sets of needles to hive a ribbed or corrugated surface to the fabric. This type of goods is found in bathing suits, underwear, sweaters, scarves and knitted caps.

■ ribbon

a fillet or narrow woven fabric of varying widths, commonly one-quarter to three inches, having selvage edges, chiefly of rayon, silk or velvet and used for braiding, decoration, trimmings, etc.

■ ribbon yarns

yarns that are woven or knitted in the form of a ribbon.

■ rickrack

flat braid in a zigzag formation. Made from several types of fibres, it is much used for many kinds of trimming on apparel.

■ ring spun

a spinning system in which twist is inserted in a yarn by using a revolving traveller. This method gives a tighter twist than the more modern, faster and usually cheaper open end spinning process.

■ rippler

the coarse comb used for removing seeds from flax fibre. The next step would be 'hackling'. The preceding step would be 'scratching'.

■ rippling

the process in flax production that removes the seedpods. The tool used is a 'rippler'. After this step of flax fibre preparation, the fibres are 'retted'.

■ rip-stop nylon

a lightweight, wind resistant and water resistant plain weave fabric. Large rib yarns stop tears without adding excess weight to active sportswear apparel and outdoor equipment such as sleeping bags and tents.

■ riser

structure which holds a pipe that conveys gas or oil from a well to a drilling platform. The riser extends from the sea floor: where it protects the well from seawater to the platform.

■ robust wool

wool possessing a strong hand and bulky nature.

■ rolag

the cigar-shaped roll of carded fibre, loosely rolled off of the hand cards, used as the fibre source when spinning woollen yarn.

■ **rooed**

the process where fleece is plucked off of Shetland sheep during the spring. This works with Shetland fleece as the sheep produce a weak spot earlier in the season.

■ **rotor spinning**

a method of open end spinning which uses a rotor (a high speed centrifuge) to collect and twist individual fibres into a yarn.

■ **roving**

the soft strand of carded fibre that has been twisted, attenuated and freed of foreign matter preparatory to spinning.

■ **RTM (composites)**

Resin Transfer Moulding. It allows the moulding of components which have complex shapes and large surface areas with a good surface finish on both sides. The process is suited to short and medium runs and is employed in many transport applications such as truck cabs. This process consists of placing reinforcements in the mould before injecting the resin. Polyesters, epoxies, phenolics and acrylics are usually used. Various kinds of moulds are used and heat may be applied to assist the cure, in which case a steel mould may be necessary. Low profile resins can be used to improve surface finish and appearance. Alternatively, low-pressure RTM allows cheaper composite tooling to be used. The reinforcement can be continuous filament mats or fabrics.

■ **rubber**

a manufactured fibre in which the fibre-forming substance is comprised of synthetic rubber.

■ **rubberised fabric**

any fabric with a rubberised coating on one or both sides making it waterproof and resistant to most stains.

■ **ruching**

adding a frill of lace or other material, often pleated.

■ **run of the loom**

fabric as it comes from the loom and ready for shipment without inspection and without elimination of defects made in weaving.

■ **run of the mill**

1. textile products of practically any type that are not inspected or do not warrant inspection.
2. fabrics of any variety that cannot be classed as a 'first'.

■ **sailcloth**

any fabric used for sails, usually a heavy and sturdy made canvas of cotton, linen, jute, polyester, or nylon. A lightweight popular fabric in use at the present time is on the order of balloon and typewriter fabric and it is much used in spinnakers and headsails. The cloth is finished around 40 inches and texture is approximately 184-square with a weight of about six yards to the pound.

■ **sandwash**

the soft peach skin finish obtained by blasting a fabric with fine sand.

■ **sanforising**

a controlled compressive shrinkage process.

■ **saran fibre**

a manufactured fibre which has an excellent resistance to sunlight and weathering and is used in lawn furniture, upholstery and carpets.

■ **sari patterns**

traditional Indian sari designs.

■ **sateen**

this cloth is made with a 5-end or an 8-shaft satin weave in warp-face or filling-face effects. Filling-face sateen requires a great many more picks, than ends per inch in the goods while the reverse is true in the warp-face material. Combed yarn sateens are usually mercerised and have a very smooth, lustrous surface effect.

■ **sateen fabric**

a fabric made from yarns with low lustre, such as cotton or other staple length fibres. The fabric has a soft, smooth hand and a gentle, subtle lustre. Sateen fabrics are often used for draperies and upholstery.

■ **sateen weave**

a variation of the satin weave, produced by floating fill yarns over warp yarns.

■ **satin**

the name originated in Zaytun, China. Satin cloths were originally of silk and simulations are now made from acetate, rayon and some of the other manmade fibres. The fabric has a very smooth, lustrous face-effect while the back of the material is dull. The satin weave and finishing of the goods provide and enhance the lustrous surface texture. Some fabric is made with a cotton backing such as cotton-back satin or crepe satin. Satin fabrics may be light or heavy in weight, soft, crepe-like, semi-stiff in finish and hand. Acetate and rayon satins are popular in the lower price range. The major types of satin are:
1. Satin Crepe: It is a soft, drapy, lustrous fabric used in dress goods, formal wear, negligees, underwear, bedspreads and draperies.
2. Panne Satin: Often known as 'Slipper Satin', the fabric is rather stiff in hand and it finds much use in evening wear, cloaks and bridal wear.
3. Lining Satin: Made of panne or crepe satin fabric it is lighter in weight when compared with the so-called dress satin. Much used in coat lining, some formal wear and dress goods.
4. Wash Satin: Can be either panne or crepe in type and is finished for laundering purposes. Chief uses are for blouses, lingerie and some dress goods.

5. Upholstery Satin: This is the 'heavy satin fabric' and is made with cotton back construction. Used in upholstery, drapes, curtains and to some degree in formal eveningwear.

6. Slipper Satin: Interchangeable with Panne Satin, but not always correctly, this type of goods is a richly embellished material interspersed with metallic threads to enhance the motif. Jacquard motifs are used in the construction and the goods run from the conservative to the bizarre.

■ satin damask

1. a heavy, rich silk cloth made on the Jacquard loom, with fancy weaves and embellishments or in a pile construction. Used for hangings and curtains.
2. the best quality of linen of damask used for table linen.

■ satin fabric

a traditional fabric utilising a satin weave construction to achieve a lustrous fabric surface. Satin is a traditional fabric for evening and wedding garments. Typical examples of satin weave fabrics include slipper satin, crepe-back satin, faille satin, bridal satin, moleskin and antique satin.

■ satin weave

one of the three basic weaves, the others being plain weave and the twill weave. All other constructions, plain or fancy, must be made from these weaves in variations, either alone or in combination. The surface of satin weave cloth is almost made up entirely of warp or filling floats since in the repeat of the weave each yarn of the one system passes or floats over or under all but one yarn of the opposite yarn system. Intersection points do not fall in a straight line as in twills but are separated from one another in a regular or irregular formation. Satin weaves have a host of uses-brocade, damask, other decorative materials, many types of dress goods, formal and evening wear apparel, etc.

■ saturation

also known as chroma, intensity and purity. It is the strength or purity of a colour, intense or bright, subdued or greyed. If the colour is as brilliant as possible, it is known as one of saturation or strength. If subdued or greyed, it is dull, weak and low in intensity.

■ saxony wheel

three-legged 'standard wheel' featuring a side-by-side arrangement with the flyer to the left of the drive wheel.

■ scaffold

a temporary platform used for tissue growth.

scales

cuticle cells form a scale-like formation on the surface of the fibre, resembling shingles on a roof. These scales on the surface of the fibre open from base to tip, causing an interlocking or felting action when fibres are randomly mixed during processing.

scotch tension

a single band drives the flyer. The bobbin has an adjustable friction band to slow it. When tension on the yarn is released, the bobbin rotations is stopped by the break band and the flyer winds the yarn onto the bobbin. Because this only involves one simple adjustment, many 'beginner' wheels use scotch tension.

scotch tweed

made on a two-up and two down twill with white warp and stock-dyed filling or vice-versa. The stock colours are usually rather vivid in order to give contrast in the fabric. Fibre staple in the yarn is usually variable and is irregular in appearance, often this shagginess seems to add to the looks of the material. Always popular, the cloth is used in suiting, top-coating, sport coating and some over coating.

scotchgard

it is a fluoro chemical oil and water-repellent finish applied during finishing. Fluids form globules on the surface of treated fabrics, do not spread and can be removed easily by dabbing with tissue. Oil may be cleaned off without leaving a grease ring. The finish is odourless, colourless and harmless and the hand of the fabric is not impaired. Scotchgard is fast to washing and dry cleaning.

scouring

the process of washing or cleansing wool of grease, soil and stint in a water/soap/alkali solution. When scouring is done commercially, a normal fleece goes through 3 washings. During the scouring process, a fleece may loose up to 50% of its original weight.

screen or stencil printing

silk or nylon is used as the screen in this work. It is spread over the frame, which according to the desired design, has portions of the screen surface covered or enamelled by a coating. Covered areas will not take on the dyestuff, the open areas allow the colour to pass through the screen onto the fabric upon which the screen is set. Colour is poured into the frame shell and it applied to the fabric by means of a squeegee worked back and forth. There has to be a frame for each colour used. The method is rather expensive, yardage is limited but there is a wide variety of design in this type of work.

scrim

1. an open mesh, plain-weave cotton cloth made from carded or combed yarns in several constructions and weights for use as bunting, buckram, curtains, etc.
2. cheesecloth, when bleached and firmly sized, is known as Scrim.
3. a lightweight cotton sheer cloth made in doup or in plain weave

with single-ply yarns. It is often made with coloured checks or stripes and serves as curtaining.

■ **scroop**

the rustling sound produced when some silk is compressed.

■ **scrutching**

this is a mechanical operation which, by breaking and beating the retted flax straw, separates the textile fibres in the stem of the plant from the woody matter and the bark. The next step would be 'hackling'.

■ **sea island cotton**

finest of all cotton, very white and silk-like with staple of 1.5 inches.

■ **seasonless solids**

basic colours which do not change from season to season, including black, white and navy.

■ **second cuts**

fribs or short lengths of wool resulting from cutting wool fibres twice in careless shearing. An excessive number of second cuts decreases the average fibre length and depreciates spinning quality.

■ **secondary colours**

green, orange and violet, each of which is obtained by the mixing of two primaries.

■ **seedy wool**

wool containing numerous seeds or an appreciable amount of vegetable matter.

■ **seersucker**

a woven fabric which incorporates modification of tension control. In the production of seersucker, some of the warp yarns are held under controlled tension at all times during the weaving, while other warp yarns are in a relaxed state and tend to pucker when the filling yarns are placed. The result produces a puckered stripe effect in the fabric. Seersucker is traditionally made into summer sportswear such as shirts, trousers and informal suits.

■ **selvedge**

each side edge of a woven fabric and an actual part of the warp in the goods. Usually easily distinguishable from the body of the material, other names for it are listing, self-edge, raw edge.

■ **semi-bright wool**

grease wool that lacks brightness due to the environment under which it is produced, though it is white after scouring.

■ **semi-worsted yarn**

yarn spun from sliver carded (not combed) and pin-drafted on worsted spinning system machines.

sensitised

refers to fabrics that have been impregnated with certain finishing chemicals, then dried but not fully cured. The fabric is then said to be in a sensitised condition until the proper degree of heat is applied such as that observed in the Deferred Cure process for Durable Press.

separation (geotextiles)

the function of a geotextile as a partition between two dissimilar geotechnical materials, e.g. soil and gravel. The geotextile prevents intermixing of the two materials throughout the design life of the structure.

serge

a fabric with a smooth hand that is created by a two-up, two-down twill weave.

serging

an overcastting technique done on the cut edge of a fabric to prevent ravelling.

sericin

also known as silk gum. The gummy material holding the silk filaments together.

serrations

the outer or epidermal scaly edges on the wool fibre which can be seen under a microscope. Usually the finer the wool the greater the number of serrations. Serrations assist in felting by interlocking.

sett

a term used to define the weft or warp density of a woven fabric, usually in terms of a number of threads per inch.

setting the twist

after you have plied your wool, you need to set the twist. There are several approaches to this. One school of thought says that you wash your yarn and then dry it under tension. This approach is fairly popular with weavers. Another approach says that you wash your yarn and **don't** dry it under tension. This approach is more popular with knitters. Yet another approach says to really shock your wool and let it do what it is going to do. This is done by washing in alternating hot and cold baths.

shade cloth

a plain-weave fabric that is usually white, canary, ecru or green in colour, used for shades on windows. Cloth is smooth, firm, rugged and rather lustrous in finish. Has good body and feel and the required stiffness is provided by a mixture of oil, sizes and starches. Material is not transparent and withstands rough usage.

shadow printing

chintz, cretonne, ribbon and some silks are woven with only the warp printed so as to provide mottled effects when woven with a white or light-coloured filling. These indistinct motif fabrics are reversible.

Textile

- **shafty wool**

    wool of extra good length, sound and well grown.

- **shantung**

    1. a silk fabric very similar to but heavier than pongee. Originally woven of wild silk in Shantung, China, now often made with synthetics or mixtures. Very popular for summer dresses and suits.
    2. a cotton fabric with an elongated filling yarn.

- **shape retention**

    the ability of a durable press garment to be washed and still retain the original shape of the new garment.

- **sharkskin**

    a hard-finished, low lusted, medium-weight fabric in a twill-weave construction. It is most commonly found in men's worsted suiting. However, it can also be found in a plain-weave construction of acetate, triacetate and rayon for women's sportswear.

- **shear**

    1. a shearing machine that does mechanical cutting or trimming of projecting fibres from the surface of cloth. Fabrics can be sheared to the one thirty-second part of an inch as to height of the nap on the goods.
    2. an operation in the finishing plant of a mill to shear or cut-off long floats, lappets and comparable materials.

- **shearing**

    cutting of the fleece from a sheep with hand shears or by machine power shears, now widely used. The fleece should come off the animal in a solid sheet which is then wrapped into a compact bundle with the flesh side outside.

- **sheaves**

    rollers or pulleys over which ropes, wires or umbilical may be deployed.

- **shed**

    an opening formed during weaving by raising some warp threads and lowering others to facilitate the passage of a weft yarn or a weft carrying device across the weaving machine.

- **shedding**

    a motion in weaving whereby a shed is created to facilitate the passage of a weft yarn or a weft carrying device across the weaving machine.

■ **sheepskin**
the wool still on the pelt or skin.

■ **sheer**
any of a group of very thin cloths such as chiffon, batiste, net, organdy, voile, etc. 'Heavy sheer' and 'semi-sheer' are used to describe the more compact goods in this family of fabrics made from the same fine yarns employed but with higher textures than in ordinary sheers. Used for dress goods, eveningwear, bridal wear, etc.

■ **sheeting**
plain-weave carded or combed cloth that comes in light, medium and heavy weights. Sheeting for converting purposes is usually about 40 inches wide. Sheeting comes in the following classifications-coarse, ordinary, lightweight, narrow, soft-filled and wide. It may be unbleached, semi-bleached, full-bleached or coloured. Industrial sheeting serves as backing for artificial leather, boot and shoe lining, etc.

■ **shepherd's check**
a small check effect in contrasting colours, often black and white.

■ **shetland**
a wool yarn or fabric with a soft yet firm handle, plain dyed or in mixture shades.

■ **shifu**
thread made from paper is an old Japanese tradition and has been used historically in clothing.

■ **shin gosen**
fabrics made from ultra-fine polyester filament yarns with enhanced comfort, handle, drape and aesthetics. Shin gosen fabrics are designed specifically to appeal to end users by employing a combination of sophisticated fibre and fabric processing technologies.

■ **shirring**
making puckers or gathers in a fabric, often by using elasticated thread in parallel rows.

■ **shirting textures and sizes**
1. broadcloth-100 to 136 by 60 to 64.
2. Madras-80 to 84 by 72 to 80.
3. Oxford-88 to 90 by 40 to 50.
4. common print cloth-80 square by 68 to 72. Sizes range from 13 ½ inches to 57 ½ inches, waist, from 36 ½ inches to 55 ½ inches.

■ **shives (flax)**
short pieces of woody waste beaten from flax straw during scutching.

■ **shoddy**
wool fibres recovered from either new or used woven or felted cloth and which must be designated as reprocessed or reused. Wool fibres

included in this classification usually run .5 inch or more in length.

■ **shorts**

short pieces or locks of fibre that are dropped out while fibres are being sorted.

■ **shot**

a colour effect seen in a fabric woven with a warp of one colour and a weft of a contrasting colour.

■ **shrinkage**

the reduction in width and length or both, that takes place in a fabric when it is washed or dry-cleaned. Residual shrinkage is the term used to indicate the percentage of shrinkage that occurs in the fabric at the time of its first washing. Shrinkage that may occur on each subsequent washing is the progressive shrinkage.

■ **silicone**

generic term for one of a class of organic chemical compounds, based on the partial substitution of silicon for carbon. Obtained from silicon, a component of sand, it is used as a softening agent in finishing fabrics.

■ **silk**

the only natural fibre that comes in a filament form, from 300 to 1,600 yards in length as reeled from the cocoon, cultivated or wild. When the silkworm begins spinning, two filaments are emitted form the 'silk ducts' which are covered by Silk Gum or Sericin from the sacks before they come from the mouth. As the liquid is emitted by the silkworm it solidifies on contact with the air. A single filament is called Fibroin or Silk. The two filaments joined together produce what is known as the Cocoon Thread or Bave.

■ **silk finish on cotton**

a full finished of mercerising, gassing and schreinering. Glauber salts in the sizing bath add lustre when correctly used. A rustling effect, as in silk taffeta, is noted in cloth finished in this manner. Oil softeners are often used to make the goods softer. The finish may be given to cotton sateen, mescaline and charmers.

■ **silk floss**

1. very short fibres of tangled waste silk.
2. erroneously used to describe the soft fibres from the kapok tree which are blended together and used as stuffing for pillows, mattresses and life preservers.

■ silk noil

it is a by-product of the spun-silk industry. It consists of short fibres which are combed our of the silk waste and are not suitable for clean, even yarns. Silk noil, nevertheless, is spun into noil yarns, which are very uneven and lumpy. It is also used in blends for novelty effects.

■ simulated linen finish

applied to cotton fabric, it is obtained by beetling to give a soft, full, kidskin type of 'hand'. Mercerising is also given the cloth that requires special sizing and pressings so as to flatten out the fabric to produce the effect. Rays of light are reflected by the finish, which is not permanent and often not durable.

■ single damask

silk fabric which has the ground and the motif weave or weaves made on a five-shaft satin weave. The double damask is made on the eight-harness satin weave and would not have the short floats noted in the single damask construction which gives better service. Also made in linen, rayon and mixture fabrics.

■ single knit

a fabric knitted on a single needle machine. This fabric has less body, substance and stability when compared with double knit.

■ single yarn

one that has not been plied, the result of drawing, twisting and winding a mass of fibres into a coherent yarn.

■ singles

the individual unit of yarn. Referring to a 'single ply' is almost guaranteed to make experienced spinners cringe.

■ sintering

a process in which larger particles are formed by applying heat and/or pressure to a powder.

■ sirospun yarns

worsted ply yarns spun on a slightly modified ring-spinning frame, which creates the yarns directly from two roving. In forming the yarns, the spinning frame twists the two roving together, thereby holding the fibres in place. The process, developed in Australia, eliminates the step of forming two separate single yarns.

■ sisal

a vegetable fibre that is made into strong, coarse twine. It is used for binder twine, but should not be used to tie fleeces.

■ size

any of various gelatinous or glutinous preparations made form glue, starch, etc., used for coating the threads.

■ sizing

1. application of a size mixture to warp yarn, the purpose of which is to make the yarn smoother and stronger to withstand the strain of weaving, to provide an acceptable hand in the woven grey goods and to increase fabric weight.
2. application of starch or other stiffeners to fabrics, garments, etc.
3. the process of determining the count or number of roving or yarn. Skein a length of yarn taken from the reel.

■ skein winder

a traditional, low-tech way to wind a skein and measure yarn.

■ skeining

the processing of winding a skein of yarn.

■ skirting

the practice of removing from the edges of the whole fleece at shearing time of all stained and inferior parts.

■ sleazy

said of a fabric when it lacks firmness, is poorly woven, is not textured correctly, has an unappealing hand and is poorly finished by being too smooth, slippery and generally speaking, a very inferior piece of goods.

■ sliver

the strand of loose, untwisted fibres produced in carding.

■ slub yarn

yarn of any type which is irregular in diameter, may be caused by error or purposely made with slub to bring out some desired effect to enhance a material. Slub yarns are popular as novelty threads in summer dress goods and find much use in hand-woven fabrics.

■ SMC

Sheet Moulding Compound.

SMS

a non-woven structure consisting of Spunbond, Meltblown, Spunbond layers.

snapping

a method of testing the individual locks of raw wool. This is done by holding the two ends of a lock of wool and pulling your hands quickly apart. There should be an audible snapping sound—but no damage to the lock. This is also done, on a very different scale, with a just-washed skein of yarn. For this situation, put your arms through a skein and 'pop' the skein. This will even out most kinks in the skein.

snarls

small, curly or 'kinked' places in yarns.

soft goods

term sometimes applied to textile fabrics.

softener

any of a large number of chemical compounds used in fabric finishing to give the cloth a mellow, soft and appealing hand or handle.

softening

application of a soft finish on goods without adding a sleazy or clammy feel.

soil release

the property of a fabric permitting removal of most oil and waterborne stains by ordinary home laundering. Suggests a special finish. Dual action Scotchgard fabric protector imparts soil release to many permanent press fabrics.

soil release finish

refers to one of several finishes used on durable press blends which provides for greater ease in cleaning the article. Some soil release finishes also provide resistance to soiling as well as ease of soil removal.

soil retardants

various chemical compounds that are applied to fabrics, especially carpets, to enable them to resist soiling.

soleiado

a term, originally the name of a company, used to describe a Provencal print.

solution dyeing

also called Dope Dyeing and Spun Dyeing, the pigment or colour is bonded-in in the solution and is picked up as the filaments are being formed in the liquor. Cellulose and non-cellulose fibres are dyed to perfection by this method. Colours are bright, clear and fast.

sorting

the process of separating a fleece into its various qualities according to diameter, length, colour, strength and other factors.

sound wool

wool that has a strong staple. Wool buyers or graders test the soundness of the wool by holding a staple at either end and snapping their fingers across the middle of it.

space dyeing

a dyeing process in which yarn is coloured at intervals.

spaced dyed yarn

yarn dyed in single colour or multicolour spaces along a given lineal length or yarn in either repeat type or random type patterns.

spacer fabric

three-dimensional structures consisting of two warp or weft-knitted layers connected by monofilament spacer yarns. They can also be knitted on double jersey circular or on electronic flat machines.

spandew

a manufactured fibre in which the fibre-forming substance is composed of a long-chain polymer of at least 85 per cent of a segmented polyurethane.

spandex fibre

a manufactured electrometric fibre that can be repeatedly stretched over 500% without breaking and will still recover to its original length.

specking

the removal of specks, burrs and other detrimental objects that might impair the final appearance of woollens and worsteds. This is usually done with tweezers or burling irons.

spin drawing

a process for spinning partially or highly oriented filaments in which the spinning and drawing processes are integrated sequential stages. Most of the orientation in spin drawing is introduced between the first forwarding device and the take-up.

spindle

1. a slender, tapered steel rod set in vertical position along the side of a ring spinning frame upon which a bobbin revolves to receive yarn as it is spun. Also used on sliver, slubbing, roving, twisting frames etc.
2. a rounded wooden rod tapering at each end made to revolve and twist into yarn with fibres drawn out from a bunch of wool, cotton, flax, etc. in the making of hand spun yarn.
3. the amount of yarn that can be prepared on a spindle at one time. Thus, a measure of the quantity on length of yarn varying according to the material used. Also spelled as spyndle.

blue flame with silver sparkles velour

- **spindle spun**

  a yarn produced on a hand spindle.

- **spinlaying**

  part of a production route for making non-woven in which synthetic filaments are extruded and gathered on to an endless belt.

- **spinnerette**

  a rounded metallic cap or jet used in the manufacture of manmade filaments. There are one or more holes (orifices) in the flat surface or top through which the spinning solution is forced emerging into a coagulating medium. Orifices, generally speaking, range from 0.002 to 0.005 inch in diameter and there is a great variance in the number of orifices used depending on the size denier filament desires. Spinnerettes are made from iridium and platinum although other metals are now used to some degree.

- **spinner's type**

  a fleece that is strong, regular, of good colour and character and nearly free of vegetation and dirt.

- **spinning**

  this final operation in yarn manufacture consists of the drawing, twisting and the winding of the newly spun yarn onto a device such as a bobbin, spindle, cop, tube, cheese, etc. Spinning requires great care by all operatives involved. Mule and ring spinning are the two major methods today and in addition to being spun on these methods, worsted yarn is also spun on the cap and flyer frame methods of producing finished spun yarn.

- **spinning count**

  the fineness of which a yarn may be spun. The number of hanks of 560 yards. each in length to 1 pound of top. Thus, 1 lb. of fine top that will spin 64 hanks is called 64's.

- **spinning frame**

  a machine for drawing out cotton or other fibres to their final spun size, twisting them to impart strength and winding the yarn onto bobbins.

- **spinning jenny**

  an early spinning machine having more than one spindle, enabling a person to spin a number of yarns simultaneously.

- **spinning solution**

  a solution of fibre-forming polymer ready for extrusion through a spinneret.

- **spinning wheel**

  a device used for spinning fibre into yarn or thread, consisting essentially of a single spindle drive by a large wheel or a flyer unit driven by a treadle.

■ spinning, core

a spinning method wherein a base or core yarn is encased by a group of staple fibres, usually in a spiral formation. On removal of tension the staple or outside fibres are pulled into a more compact formation around the core yarn used. In stretch fabric yarn spinning, the yarn will become stretchable to the degree or extent of the predetermined tension of the elastic core filament used.

■ spinning, intimate blend

cut spandex fibres blended with tow or cut staple fibres to produce a stretch yarn at the fibre producer level.

■ spinster

a spinner whose occupation is spinning. Updated to be politically correct.

■ spiral yarn

it is made of two yarns whose yarn counts vary to a marked degree. The fine yarn has been given hard twist, the bulky yarn has received slack twist, the heavier yarn in twisting is wound spirally about the fine, hard-twisted yarn which is sometimes referred to as core thread. Other names for this yarn are corkscrew and eccentric.

■ sponging

a pre-shrinkage by dampening with a sponge, by rolling in moist muslin or by steaming, given to woollens and worsteds by the clothing maker before cutting to insure against a contraction of the material in the garment. The very popular sponging treatment is 'London shrunk' which is a cold-water treatment, originating abroad and is frequently applied and guaranteed by the cloth manufacturers themselves.

■ spooling

packaging or warp yarn onto a cheese or similar device for further manipulation.

■ spot and stain resistant

material that has been treated to resist spots and stains, there are several good compounds on the market for this purpose.

■ spot weave

a woven construction in which patterns are built in at spaced intervals through the use of extra warp and/or extra fill yarns are placed in selected areas. These yarns are woven into the fabric by means of a dobby or Jacquard attachment.

■ spring wool

the 67 months of wool produced by sheep shorn in the spring following fall shearing.

■ spun silk

a yarn composed of fibres of silk which have not been reeled from the cocoon, but have produced by piercing the cocoon and then shredding it into lengths of 3 to 15 inches, often from tussah cocoons.

■ **spun yarn**

a yarn made by taking a group of short staple fibres, which have been cut from the longer continuous filament fibres and then twisting these short staple fibres together to form a single yarn, which is then used for weaving or knitting fabrics.

■ **spunbond**

non-woven made from a continuous mat of randomly laid filaments. The filaments are bonded together by heat and pressure or needle punching.

■ **spunbonding**

the process used to manufacture spun bonded non-woven.

■ **spunlaced fabric**

a fabric manufactured by spunlacing.

■ **spunlacing**

a process for bonding a non-woven fabric by using high-pressure water jets to intermingle the fibres.

■ **spunlaid fabric**

a fabric produced by laying freshly formed synthetic filaments into a web.

■ **spunmelt**

a non-woven structure made by extruding molten polymer through spinnerets to form fibres. Spunmelt processes are used in the manufacture of spunbond non-woven, melt blown non-woven and combinations of the two.

■ **sputtering**

a process in which atoms, ions and molecules are ejected from the surface of a target material when it is irradiated by an ion beam. One application of sputtering is to exploit the conditions in which the ejected particles re-form on another substrate as a thin film or coating. For instance, thin metallic films are often applied in this way to electrically non-conductive substrates to give them conductive properties.

■ **square cloth**

a term used for any cloth having the same number of ends as picks per inch, with its warp counts the same as the filling as 80 x 80, 64 x 64.

■ **squirrel-cage swift**

this swift has two rotating cylinders mounted on a vertical post. The cylinders can be shifted to adjust for different-sized skeins.

■ **SSMMSS**

a non-woven structure consisting of Spunbond, Spunbond,

Meltblown, Meltblown, Spunbond, Spunbond layers.

■ **stability**

that property of a bonded fabric which prevents sagging, slipping or stretching. This is conducive to ease of handling in manufacturing and helps to keep its shape in wear, dry-cleaning and washing.

■ **stain resistance**

the ability of a fabric to withstand permanent discolouration by the action of liquids. This property depends partly upon the chemical nature of the fibre but may be improved by proprietary treatments.

■ **stained wool**

wool that has become discoloured by urine, dung or whatever, which will not scour out white. Badly stained pieces should be removed at shearing before the fleeces are packed.

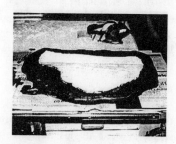

■ **staple fibres**

short fibres, typically ranging from $1/2$ inch up to 18 inches long. Wool, cotton and flax exist only as staple fibres. Manufactured staple fibres are cut to a specific length from the continuous filament fibre. Usually the staple fibre is cut in lengths ranging from $1 1/2$ inches to 18 inches long. A group of staple fibres are twisted together to form a yarn, which is then woven or knit into fabrics.

■ **staple length**

the fibre length from a sample of fibres. A Wensleydale sheep has a standard staple length of about 12 inches. A Dorset has a staple length 2.5 and 4 inches.

■ **staple length**

the length of sheared locks obtained by measuring the natural staple without stretching or disturbing the crimp. The fibre regrowth or regeneration from one shearing to the next.

■ **stent**

a narrow tube commonly used to keep blood vessels open in the arteries.

■ **stipple printing**

printing of small dotted effects set in among spaces or bare areas of a printed motif. Used chiefly in novelty effects.

■ **stitchbonded fabric**

a fabric made by stitchbonding.

stitchbonding

a process in which a series of interloped stitches are inserted along the length of a pre-formed fabric, an array of cross-laid yarns or a fibre web. Proprietary systems include Arachne, Malipol and Maliwatt.

stock dyeing

'dyed-in-the-wool' is another term for stock dyeing because the fibre is still in the form of loose fleece. Most stock-dyed wool is made into woollen yarns. Worsted wool is never dyed until after it is combed and dyeing at this stage is called top dyeing. Both methods produce the highest penetration of dye and the highest quality of coloured yarns. Wool dyed this way is also used to make speckled yarns of mixed colours and subtle heathers.

storage bobbins

bobbins used to store the yarn for subsequent plying. Depending on your school of thought, using a storage bobbin allows you more evenly wind on (then when you were spinning). This means that the plying process should also be more even.

strain

the change in length per unit length of a material in any given direction.

strength

this refers to the how much weight the fibres can bear. Some fibres, like flax, actually get stronger when wet.

stretching

this is when the fibres are pulled taut while spinning.

striated

an effect applied to a yarn to give the appearance of striations lines of colour or fine parallel scratches or grooves, as on the surface of a rock over which a glacier has flowed.

strick

the bundle of prepared (hackled) flax fibres.

stubble shearing

the practice of shearing or cutting a portion of the wool at varying lengths, from sheep used for show purposes.

s-twist

spinning clockwise. Traditionally, this is the direction 'singles' are spun. If your singles have been spun S-twist, you would ply Z-twist and then cable S-twist.

style

the combination of crimp and crinkle ranging from good crimp and good crinkle to no crimp and no crinkle.

subcontracting

an arrangement whereby one business (subcontractor) manufactures all or part of a specific product on behalf of another business (main contractor) in accordance with plans and technical specifications supplied by the main contractor. The main contractor has final economic responsibility in such an arrangement.

sub grade intrusion (geotextiles)

localised penetration of a soft cohesive sub grade and resulting displacement of the sub grade into a cohesionless material.

sublimation

a process in which a substance is changed directly from a solid into a gas or vapour without first melting.

substantively

the attraction between a fibre and a substance (such as a chemical finish) under conditions whereby the substance is selectively extracted by the fibre from the application medium (for example, water).

suede fabric

1. woven or knitted cloth made from the major textile fibres and finished to resemble suede leather. Used for cleaning cloths, gloves, linings, sport coats, etc.
2. some sheeting may be napped on the one side to simulate suede leather.

suint

generally referred to as the perspiration of sheep and is naturally excreted from the glands at the roots of the wool. Suint consists of soapy compounds of potash and fatty acids, together with a little free fatty acid and saline matter. It is soluble in water. This is one part of what is called the grease on an unwashed fleece.

sulphur dyes

they do not provide bright shades to any marked degree and their fastness properties have to be developed in the chemical inertness and insolubility in water. There are now also soluble forms of these dyes on the market. Used to colour heavy cottons, knitwear in medium to full shades and readily dye stock, yarn and piece goods. Weak in sunlight except for deep shades where fastness is good. Fugitive to chlorine-base bleaches.

superconductor

a material that can conduct electricity or can transport electrons from one atom to another with no resistance usually at temperatures near absolute zero.

superfine wool

superfine wool-from about 15 to 18 microns-is in a class by itself, comparable to fine cashmere and is used to make fabrics of the highest quality. Superfine wool comes from strains of Merino sheep that have been developed to produce especially fine fibres.

surah

a soft, twill-woven silk or rayon, often made in plaid effects. Surah prints are popular at times. If made of some fibre other than silk, the fibre content must be declared. Uses include neckwear, mufflers, blouses and dress goods. Named for Surah, India.

survivability (geotextiles)

the ability of a geotextile to perform its intended function without undergoing degradation.

swatch

in the trade, any small sample of material is often referred to by this name. Swatches are used for inspection, comparison, construction, colour, finish and sales purpose.

swift

a swift has an expanding core that can be adjusted to fit various skeins. This allows spinners to help keep the hank of yarn in some kind of order while unwinding it.

swiss

a fine, sheer cotton fabric which may be plain, dotted or figured. This crisp, stiff cloth is often called dotted swiss. Used for dress goods, wedding apparel, curtains, etc. The term 'swiss' in textiles as applied to a fabric logically and properly identifies fabrics made in Switzerland. The domestic fabric is known as domestic dotted swiss, etc.

swiss batiste

a sheer, opaque fabric noted for its high lustre which is accompanied by special finishing and the use of special grades of long-staple cotton and swiss mercerisation.

syndiotactic

polymer which has alternating stereo chemical configurations of the groups on successive carbon atoms in the chain.

synthetic dye

a complex colourant derived from coal tar.

synthetic fibres

man-made fibres made from a polymer that has been produced artificially, in contrast to fibres made from naturally occurring polymers such as cellulose. The term synthetic fibres is also used to refer to synthetic filaments.

synthetic filaments

man-made filaments made from a polymer that has been produced artificially, in contrast to filaments made from naturally occurring polymers such as cellulose.

■ **table (bench)**
chunk of wood with legs attached to bottom and everything else attached to top.

■ **taffeta**
a fine plain-weave fabric smooth on both sides, usually with a sheen on its surface. Named for Persian fabric taftan. May be solid coloured or printed or woven in such a way that the colours seem 'changeable'. Used for dresses, blouses, suits. Originally of silk, now often made of synthetic fibres. There are several taffeta classifications, such as:
1. Antique Taffeta: A plain weave, stiff finished fabric of douppioni silk or synthetic fibres made to resemble the beautiful fabrics of the 18th century. When yarn dyed with two colours, an iridescent effect is produced.
2. Faille Taffete: Taffeta woven with a pronounced cross-wise rib.
3. Moire Taffeta: Rayon or silk taffeta fabric with moiré pattern. In acetate, moiré can be fused so it is permanent.
4. Paper Tafeta: Lightweight taffeta treated to have a crisp paper-like finish.
5. Pigmented Taffeta: Taffeta woven with pigmented yarns which give the fabric a dull finished surface.
6. Tissue Taffeta: Very light-weight transparent taffeta.

■ **tag locks**
large locks of britch wool clotted with dung and dirt.

■ **tagging**
the practice of cutting the dung locks off sheep. Usually this operation is done immediately prior to shearing and it may be done prior to lambing.

■ **tags**
broken or dung-covered wool and other wastes that are swept from the floor of shearing areas.

■ **tahkli**
a small, metal-whorl supported spindle.

■ **tambour**
1. a term which signifies work being done on the embroidery machine in which the tambour stitch has been used. This stitch produces a pattern of straight ridges which cross each other in every direction at right or acute angles.
2. tambour is also a variety of Limerick lace that is made in Ireland.
3. embroidery fame, round.

■ **tanquis**
a type of long staple fibre cotton.

■ **tape yarn**
a yarn used for knitwear in the form of a tape with a large width-to-thickness ratio. Such yarns are typically formed by weaving or knitting. Knitted tape yarns are often made on circular knitting machines, giving them a tubular cross-section.

tapestry

a closely woven figured fabric with a compound structure in which a pattern is developed by the use of coloured yarns in the warp or in the weft or both. A fine binder warp and weft may be incorporated. The fabric is woven on jacquard looms and is normally used for upholstery.

tartan

wool worsted or cotton cloth made in plain weave or in a two-up and two-down twill weave. Associated with Scottish clans, the fabric originated in Spain and was called tiritana. This multi-coloured fabric may be conventional or bizarre when made in variations of colour effects. The Scottish knit is known to everyone. Other uses include blankets, cap cloth, dress goods, neckwear, ribbon, shirting, sports coats, trews or trousers.

taupe

a brownish-grey colour, from the French word for 'mole'.

tear

percentage of tops to noils. A 41 tear would refer to wool that had 20% waste.

tear resistance

a measurement of fabric strength. Also, a property imparted by using 'ripstop' yarns in close woven fabrics.

tear strength

the force necessary to tear a fabric, usually expressed in pounds or in grams.

tear strength (geotextiles)

the force required to start or continue or propagate a tear in a geotextile under specified conditions.

technical textiles

textile materials and products manufactured primarily for their technical performance and functional properties rather than their aesthetic or decorative characteristics. End uses include aerospace, industrial, marine, medical, military, safety and transport textiles and geotextiles.

tenacity

a unit used to measure the strength of a fibre or yarn, usually calculated by dividing the breaking force by the linear density.

tencel

a new fibre created from the wood pulp of specially selected trees, processed in a non-chemical, environmentally-safe way. Tencel was introduced to the world of apparel in 1992 and is the first new fibre introduction in over thirty years. The characteristics of the fibre are a subtle lustre, high-wash stability, extremely low shrinkage and good tear resistance.

tender

wool that is weak at one or more places along its length.

tendon

a tough band of tissue which connects muscle to bones.

tensile strength

the maximum load per unit of the original cross-section area obtained prior to rupture. It is the actual number of pounds resistance that a fabric will give to a breaking machine before the material is broken on the testing apparatus and may no longer be classed as a cloth or fabric.

tension control weave

a type of decorative weave, characterised by a puckered effect which occurs because the tension in the warp yarns is intentionally varied before the filling yarns are placed in the fabric.

tentering

tentering is used to straighten the fabric and dry it. The fabric is stretched taut while steaming or drying and held in place with clips or pins, called tenterhooks. (Giving rise to that slang expression of suspense, 'on tenterhooks'.)

termination

device used at the end of a rope to secure it to a vessel, anchor, buoy, structure, etc. or to join two lengths of rope. A knot is the simplest form, but greater efficiency is achieved with splices, resin sockets, or mechanical grips.

territory wool

a designation originally given to wools originating in regions west of the Missouri River. Now applies to western range wools, not including Texas and California.

terry cloth

this cloth has uncut loops on both sides of the fabric. Woven on a dobby loom with Terry arrangement, various sizes of yarns are used in the construction. Terry is also made on a Jacquard loom to form interesting motifs. It may be yarn-dyed in different colours to form attractive patterns. It is bleached, piece-dyed and even printed for beachwear and bathrobes, etc. Also called Turkish towelling.

terry velour

a pile weave cotton fabric with an uncut pile on one side and a cut pile on the reverse side. Terry velour is valued for its soft, luxurious hand. Typical uses include towels, robes and apparel.

tex

a unit of weight indicating the fineness of yarns and equal to a yarn weighting one gram per each 1000 metres.

texture

the first meaning is the actual number of warp threads and filling picks per inch in any cloth that has been woven. It is written, say 88 x 72. This means that there are 88 ends and 72 picks per inch in the fabric. When texture is the same, such as 64 x 64, the cloth is classed as 'square' material. Sheeting, with regard to texture, is often referred to, for example, as Number 128, Number 140, etc. Consideration of the number 128 means that the total number of ends and picks per inch is 128. Thus, there might be in the texture 72 ends and 56 picks or 68 ends and 60 picks. A 140 sheeting would be better than a 128 sheeting since there would be more ends and picks to the inch in the former. Texture is also much used by the public and in advertising circles to mean the finish and appearance of cloth for sale over the counter or in the finished garment state.

■ **textured yarn**

a continuous filament yarn that has been processed to introduce durable crimps, coils, loops or other fine distortions along the lengths of the filaments.

■ **texturing**

a process during which a textured yarn is produced.

■ **texturising**

combining the strong performance properties of continuous filament nylon yarn with stretch, bulkiness and much improved absorbency and hand, through the introduction of crimps, loops, coils or crinkles into an otherwise smooth continuous filament. Texturising can be done by any one of several highly ingenious techniques known as Agilon, Fluflon, Conventional Helanca, Mylast, Saaba, Spunize, Superloft, Taslan, Textralised and A.R.C.T. Inc. in textured yarns. Except for the Taslan Method, all these processes depend on the thermo plasticity of nylon and its ability to be deformed, heat-set and developed.

■ **thai silk**

silk from Thailand typified by a rough texture.

■ **thermoplastic bonding**

a thermoplastic fibre with a lower melting point than the base fibre is blended with the latter in the formation of the web. The web is either hot-calendared or embossed, at the softening point of the thermoplastic fibre, thereby causing bonding to take place. The bonding agent constitutes from 10 to 30 per cent of the weight of the finished fabric.

■ **thermoplastic yarns**

yarns which are deformable by applying heat and pressure without any accompanying change. The deformation is reversible.

■ **thermoplastics**

a type of resin or polymer which can be remelted after cross-linking. Examples include polyolefins, such as polyethylene and polypropylene, polyvinyl chloride (PVC) and polyethylene terephthalate.

■ **thermosets**

thermosetting resins or polymers formed by chemical cross-linking which renders them permanently solid. This reaction is irreversible and, unlike thermoplastics, thermosets do not melt when heated.

Typical thermosets are polyesters, acrylics, epoxies, phenolics and vinyl esters.

■ **thigh-spun**

a yarn produced by aboriginal people.

■ **thread**

thread is made from yarn but yarn is not made from thread. It is a highly specialised type of yarn used for some definite purpose such as sewing, basting and embroidery work. Thread is plied to give it added strength when it is being manipulated. Three-ply and six-ply thread are two of the common threads in use today.

■ **thread count**

1. the actual number of warp ends and filling picks per inch in a woven cloth. Texture is another name for this term.
2. in knitted fabric, thread count implies the number of bales or ribs and the courses per inch.

■ **three-cord thread**

the plying of three single cotton yarns into one yarn. Twist inserted is always in the same direction as the spinning frame twist.

■ **through-air bonding**

a process in which a web containing fibres with a low melting point is bonded in a carefully controlled hot air stream.

■ **ticking**

compactly woven cotton cloth used for containers, covers for mattresses and pillows, sportswear, (hickory stripes), institution fabric and work clothes. It is a striped cloth, usually white background with blue or brown stripes in the motif.

■ **tie-dye**

a traditional dyeing process in which fabric is tied and dyed.

■ **tippy wool**

staples which are encrusted with wool grease and dirt at the weather end.

■ **tissue**

1. the lightweight versions of fabrics such as batiste, chambray, crepe, dimity, faille, gingham, organdy, taffeta, voile, etc. are known by this term.
2. curtains with clipspot motifs are also called tissue.
3. damask, brocade, brocatelle and some other Jacquard cloths in

which metallic threads are interspersed for enhancement of the goods use this term well.

■ **titre**
linear density.

■ **tog**
a unit used in Europe for the insulating properties of items such as duvets and sleeping bags. It is defined in the British Standards BS47451990 and BS53351991. Tog is analogous to the US clo unit (1 tog = 0.64 clo).

■ **toile de jouy**
classic designs originally created in the 1760s for the French court by textile designers in the town of Jouy en Josas.

■ **tonne**
1,000 kilograms.

■ **top dyed**
often referred to as Vigoureux Printing, it is the dyeing or printing of worsted top or silver in a rather loose formation of combed, parallel fibres. Precedes the spinning of the yarn and affords a host of colours, cast and shades.

■ **top**
sliver which forms the starting material for the worsted and other drawing systems. Tops are usually formed by combing or by the cutting or controlled breaking of continuous filament man-made fibres and the assembly of the resultant staple fibres into sliver.

■ **torque**
a force which tends to cause rotation, usually due to twist having been inserted into a yarn or removed from a yarn.

■ **total fleece weight**
the weight of the entire raw fleece.

■ **tourmaline**
a group of hard, glassy minerals used in optical and electrical equipment and in jewellery.

■ **tow**
the name given to an untwisted assembly of a large number of filaments, tows are cut up to produce staple fibres.

■ **tow linen**
these are the shorter, less desirable flax fibres separated from baste line fibres in hackling. Tow linen is usually carded and spun into a woollen-style yarn. A wonderful use for tow linen is to knit bath mitts - a situation where you want all of the rough, scratch nature of this yarn.

■ **towelling**
general term for bird's eye, crash, damask, glass, honeycomb, huck, huckback, twill, Turkish or terry, fancy, novelty and guest towelling. Many of these cloths have coloured or fancy borders or edges, some of them are often union fabrics. All towelling has property of good absorption.

■ **TPI**
Twists Per Inch (or Turns Per Inch).

■ **transmissivity (geotextiles)**
a measure of the ability of a geotextile to transmit fluids within its plane.

■ **trauma**
injury to a living tissue.

■ **treadle**
flat bit under foot.

■ **tree bark**
a ripple or wavy effect which sometimes appears on a bonded fabric only when it is stretched in the horizontal direction or widthwise. It is caused by bias tensions which happen when two distorted or skewed fabrics are bonded.

■ **triacetate**
a manufactured fibre, which like acetate, is made by modifying cellulose. However, even more acetate groups have been added to create this fibre. Triacetate is less absorbent and less sensitive to high temperatures than acetate. It can be hand or machine washed and tumble dried, with relatively good wrinkle recovery.

■ **tricot**
1. a type of warp knitted fabric which has a thin texture since it is made from very thin texture since it is made from very fine yarn. The fabric is made on one, two, or three bar frames. It is knitted flat and made on spring-beard needles and has from one to four warps or thread systems which are mounted in a stationary position.
2. stock-net as applied to a warp-knitted fabric irrespective of the motif, often refers to a flat knitted cloth since it is not tubular. The meaning, however, is not to be construed to imply a flat-machine knit fabric.
3. a French serge lining fabric made on a 20-inch width.
4. a fine woven worsted made on the tricot weave which presents fine break lines in the filling direction. This chain-break effect fabric is dyed all popular shades, has high, compact texture and is a good material to use in tailoring. Gives excellent wear in the better type of tailored garments for women.

■ **tricot knit**
a warp knit fabric in which the fabric is formed by interloping adjacent parallel yarns. The warp beam holds thousands of yards of yarns in a parallel arrangement and these yarns are fed into the knitting area simultaneously. Sufficient yarns to produce the final fabric width and length are on the beam. Tricot knits are frequently used in women's lingerie items such as slips, bras, panties and nightgowns.

■ **tricot warp knitting machine**
a warp knitting machine using bearded or compound needles mounted vertically, or nearly so, in which the fabric is supported and controlled by sinkers. The fabric is removed from the knitting point at approximately 90° to the needles' movement.

■ **tricot, warp knitted**
a warp knitted fabric knitted with two full sets of warp threads, each

set making a 1 and 1 lapping movement but in opposite directions. Additionally the term is now used generically to cover all types of warp knitted fabric made on tricot warp knitting machines.

■ **tricotine**

a weft-face woven fabric, originally with a cotton warp and worsted weft, which displays a fine, flat twill line.

■ **trilobal**

a fibre with a three-pointed, star-shaped cross-section. This gives the fibre rigidity and resilience. Also, it has many reflecting surfaces which are efficient at scattering light to hide dirt. For these reasons, trilobal fibres are often used in carpets. The reflective surfaces can also give the fibre a sparkling appearance.

■ **true-to-type wool**

a fleece showing strong breed-specific characteristics.

■ **tuck stitch**

a stitch consisting of a held loop.

■ **tufted fabric**

a fabric decorated with fluffy tufts of soft twist, multiple-ply cotton yarn. Some are loom-woven but the majority have the tufts inserted and cut by machine in a previously woven fabric, such as muslin sheeting, lightweight duck, etc. The tufts may be intermittently spaced giving the type called candlewick, or arranged closely in continuous lines giving the type called chenille. The patterns vary from simple line effects to elaborate designs. Used for bedspreads, robes, bathmats, stuffed toy, etc.

■ **tufting**

1. the tassel on an academic cap.
2. tufted fabrics such as seen in some bed covers, curtains and novelties.

■ **tulle**

fine, very lightweight, machine-made net in which small mesh effects, usually hexagonal in shape are seen. Uses include ballet costumes, bridal veils, dress goods and formal gowns. This cool dressy, delicate fabric is difficult to launder.

■ **tussah**

name for wild silk raised anywhere in the world. Compared with cultivated or true silk, it is more uneven, coarser and stronger and comes in shades of ecru through brown. Many fabric of this fibre are known merely as Tussah. Difficult to dye or bleach, it finds much use in pongee, shantung and is ideal for filling in pile fabrics and for use in blended fabrics in staple lengths.

tweed

a rough, irregular, soft and flexible, unfinished shaggy wool. One of the oldest and most popular outerwear fabrics used made of a two-and-two twill weave, right-hand or left-hand in structure. Donegal, often called tweed, is actually a homespun cloth since it is made from the plain cloth, all types of coatings and ensemble, sportswear, suiting, etc.

twill weave

identified by the diagonal lines in the goods. It is one of the three basic weaves, the others being plain and satin. All weaves, either simple or elaborate, are derived from these three weaves. Most twills are 45 degrees in angle. Steep twills are made from angles of 63, 70, 75 degrees while reclining twills use angles of 27, 20 and 15 degrees. Right-hand twilled cloth include cassimere, cavalry twill, covert, elastique, gabardine, serge, tackle twill, tricotine, tweed, whip cord. Left-hand twills include denim, galatea, jean cloth, some drill and twill cloth and some ticking fabrics.

twist

a term that applies to the number of turns and the direction that two yarns are turned during the manufacturing process. The yarn twist brings the fibres close together and makes them compact. It helps the fibres adhere to one another, increasing yarn strength. The direction and amount of yarn twist helps determine appearance, performance, durability of both yarns and the subsequent fabric or textile product. Single yarns may be twisted to the right (S twist) or to the left (Z twist). Generally, woollen and worsted yarns are S-twist, while cotton and flax yarns are typically Z-twist. Twist is generally expressed as turns per inch (TPI), turns per metre (TPM), or turns per centimetre (TPC).

twist liveliness

the tendency of a yarn to twist or untwist spontaneously.

twitt

a term applied to yarn which is irregular, that is thick and thin, the thin places being below the count required and thick places above. The defect is caused by material being drafted at an irregular rate.

type

a wool class sharing set characteristics. These are based on breed, condition, length, spinning quality, soundness, style and colour.

tyre cord fabric

a fabric that forms the main carcase of a pneumatic tyre. It is constructed predominantly of a ply warp with a light weft to assist processing.

■ **tyre yarn**
yarn that is used in the manufacture of the textile carcase of rubber tyres.

■ **UHMWPE**
Ultra-High Molecular Weight Polyethylene.

■ **ULPA**
Ultra-Low Penetration Air (filtration).

■ **Ultraviolet Protection Factor (UPF)**
a measure of the amount of protection against ultraviolet radiation to human skin provided by an item of clothing. UPF is similar in concept to the sun protection factor (SPF) which is used to categorise sun creams and lotions. However, SPF is more precise.

■ **ultraviolet stability (geotextiles)**
the ability of a geotextile to retain strength upon exposure to ultraviolet light over a specified period.

■ **umbrella swift**
a swift that opens up with a mechanism rather like an umbrella's.

■ **unbleached**
many fabrics, especially cottons, in the trade come in an unbleached or natural condition. Materials of this type have a sort of 'creamy' or somewhat 'dirty' white colour cast and much foreign matter is often seen in them-burrs, neps, nubs, specks, et al. These fabrics are stronger than full-bleached fabrics. Examples are canvas, duck, unbleached muslin, Osnaburg, cretonne, sheeting and some towelling.

■ **unevenness**
with wool, this refers to a fleece that varies in type over the body.

■ **unfinished worsteds**
a woven fabric made from worsted yarn that has then been brushed. This produces a firm fabric with a tight weave hidden beneath a soft nap.

■ **union dyed**
the colouring of two or more different textile fibres in the one dye bath to provide different colours simultaneously or dyeing the fabric a single shade, usually the latter.

■ **unit production systems**
an advanced apparel manufacturing system in which a single garment is progressed through a sequence of operations. Using a unit production system, a garment is automatically transported via a computer-controlled overhead hanging system, which has been ergonomically designed to reduce the amount of handling of the garment.

■ **unwashed wool**
wool in its original condition as it comes from a sheep.

■ **up twist**
using the same direction as the preceding spinning. So if you spun your singles 'S' and then plied them 'S' you would have used up twist. You would also have a yarn that could be used in collapse fabrics.

- **upholstery velvet**
  a wide cut or uncut heavyweight velvet which appears in plain or pattern effects. Used in draperies and upholstery.

- **upright wheel**
  flyer mounted above the wheel, making the wheel more portable. Will usually fit on a car passenger seat with the seat belt holding it.

- **uptwisting**
  a system of twisting one or more yarns by withdrawing them over-end from a rotating package. Up twisting forms the second stage in two-stage twisting.

- **value**
  the relative lightness or darkness of a colour.

- **vascular graft**
  graft such as blood vessels or heart valves.

- **vat and vat-soluble dyes**
  these are the fastest dyes known to man, insoluble in water and made soluble by chemical reduction. They actually bond-in the colourant and are the most resistant of any types of any types of light, dry-cleaning, sunlight and washing. Has many applications on cotton, rayon, polyesters, etc. Vat dyes are used to colour awnings, bed linens, decorative fabrics, outerwear, sportswear, towelling and work clothes.

- **vat dyed**
  cloth dyed by the use of vat dyes, which are obtained through oxidation. Very fast in all respects. Vat dyeing may be considered to be a misnomer since fabric coloured with these dyes are piece dyed in the conventional manner.

- **vegetable fibres**
  abaca, coir, cotton, flax/linen, hemp, henequen, istle, jute, kapok, manila hemp, pineapple fibre, ramie, sisal, straw, sunn.

- **vegetable matter**
  any kind of bur, seed, chaff, grass or other vegetable matter found in your fibre source. Also referred to as 'VM'.

- **veiled wool**
  the wool where the staple lengths have become disorganised and intermixed with each other.

- **velour**
  1. a term loosely applied to cut pile cloths in general, also to fabrics with a fine raised finish.
  2. a cut pile cotton fabric comparable to fabrics with a fines raised finish.
  3. a cut pile cotton fabric comparable with cotton velvet but with a greater and denser pile.
  4. a staple, high grade woollen fabric which has a close, fine, dense,

erect and even nap which provides a soft, pleasing hand.

5. A popular knit fabric similar to woven velour in properties, especially in hand. Ideal for men's, women's and children wear.

■ **velvet**

a warp pile cloth in which a succession of rows of short cut pile stand so close together as to give an even, uniform surface, appealing in look and with soft hand. First made of all silk, many major fibres are now used in the constructions. When the pile is more than one-eighth of an inch in height the cloth is then called plush.

■ **velvet, double**

two fabrics are woven, one on top of the other, in this plan. After the cloth is woven, a set of horizontal cutting blades cut the pile warp, thereby giving two distinct materials with cut-pile effect. There are usually eighteen blades in the cutting set-up.

■ **velveteen**

a filling pile cloth in which the pile is made by cutting an extra set of filling yarns which weave in a float formation and are woven or bound into the back of the material at intervals by weaving over and under one or more warp ends.

■ **venetian fabric**

a smooth-faced satin weave wool fabric which has been cropped to reveal a fine diagonal twill.

■ **vertical lapping**

a process in which a web is fed downwards to form vertical layers alongside each other, thereby creating a corrugated structure.

■ **vicuna**

the most coveted of all specialty hair fibres, from the smallest and wildest member of the llama family. This costly, luxurious fibre is finer than merino wool, with a rich, beautiful colour that ranges from golden chestnut to cinnamon. Each animal yields only a few ounces. The very limited supply of vicuna is controlled by the Peruvian government.

■ **virgin wool**

wool that has been clipped from a live sheep and that has not been previously processed to the stage where it contains twist. Noil is merely separated from long fibres in combing and is considered virgin wool.

■ **viscose**

the most common type of rayon. It is produced in much greater quantity than cuprammonium rayon, the other commercial type.

■ **viscose fibre**

the generic name for a type of cellulose fibre obtained by the viscose process.

■ **voile**

combed yarn, high-twist cotton staple fabric also made from some other fibres at present. This thread-like appearing cloth is made from gassed yarns which range from $^2/_{100}$ to $^2/_{200}$ in yarn count. There are five types of voile-piqué, seed, shadow, stripe and splash.

waffle cloth

fabric with a characteristic honeycomb weave. When made in cotton, it is called waffle piqué. Used for coatings, draperies, dresses and towelling. Same as honeycomb cloth.

wale

1. chain loops that run lengthwise in knit fabric, course in knit cloth runs in horizontal direction.
2. ribs in knit fabric.
3. ribs or cords observed in corduroy fabric, these may be wide or when fine in the fabric they are called pinwales. These may run from 61 to 22 in number per inch in the goods.

warp

the yarns which run vertically or lengthwise in woven goods.

warp knit

a type of knitted fabric construction in which the yarns are formed into stitches in a lengthwise manner. Warp knits are generally less elastic than weft knits. Common examples of warp knits are tricot knits and raschel knits.

warp knitting

a method of making a knitted fabric in which the loops made from each of several warp threads are formed substantially along the length of the fabric. Warp knitting is characterised by the fact that each warp thread is fed more or less in line with the direction in which the fabric is produced.

warp pattern

one repeat of the different arrangement of colours of warp threads that go to make the pattern. The filling pattern is the arrangement of the filling colours and yarn used to make one repeat. Plaids, over plaids, Glens and checks are examples.

warp pile

fabric made with two sets of warp yarn and one set of filling yarn, the extra warp-yarn set usually appears on the surface of the cloth, except in the case of towelling, where the pile effect is seen on both sides. The extra pile warp may be cut or uncut. Incidentally, velvet is an example of a warp-cut-pile cloth while corduroy and velveteen are filling cut-pile materials. Terry cloth is an example of an uncut warp-pile fabric.

warp printing

comparable to the principles of printing ordinary cloth except that the mechanical operation is different. Warp yarn is printed on a beam and then it is rewound onto a second beam that is placed in the back of the loom and made ready for the weaving operation. Warp prints give a melange or mottled effect when woven with plain filling, white or in some light colour.

warp-face

a weave in which the warp yarn predominates on the face.

warping

transferring of the warp yarn from a cheese onto the warp beam which is built up in sections to make the complete warp beam filled with the yarn. Full beams may weigh as much as 1,500 pounds.

warping board or frame

a rectangular frame with strong pegs inserted into its sides used to wind a warp.

warping reel

a rotating frame that can be mounted horizontally or vertically. It is used to wind a warp and help keep the threads in order.

wash fast

a term that seems to cause considerable confusion, especially among consumers. The term is applied, rather loosely at times, to fabrics or garments which can be washed and laundered. Much depends on the properties of the article before the term should be applied. All labels or tags should be carefully read and directions for washing followed as given.

wash ability

relatively new techniques have been developed to improve wool's wash ability by making the fibre more resistant to felting and shrinkage. One process takes place before the yarn is spun. The loose fleece or combed top is chlorinated to remove the tips of the 'scales' and then coated with an extremely fine resin. The resin masks the fibre's scales and keeps them from interlocking, which is the cause of felting and shrinkage.

washable

materials that will not fade or shrink during washing or laundering. Labels should be read by the consumer to assure proper results. Do not confuse with 'wash-and-wear'.

wash and wear

applied to a garment it is one that can be washed by hand or in a washing machine at the warm water setting. In common usage, drip-dried garments do not normally retain creases or pleats but do recover sufficiently from wrinkles to need little, if any ironing. Durable press, however, does not retain creases and pleats. Washing and drying conditions should always be specified on the so-called wash-and-wear garments.

washed wool

wool washed in cold water while on the sheep's back before shearing (industrial term).

wastage

in weaving, this refers to the part of the warp that cannot be used, often about 24-36 inches of the total warp. In spinning, this refers to the loss between the weight of fibre acquired and the weight of fibre that can be used. In spinning, the amount of dirt, tag ends and unusable fibre are all part of the wastage.

- **wasty wool**

  wool that is short, weak and tangled, which often carries a high percentage of dirt or sand.

- **water repellence**

  the ability of a fabric to shed water to a limited degree.

- **water repellent**

  ability of a fabric to resist penetration by water, under certain conditions. Various types of tests are used and these are conducted on samples before and after subjection to standard washing and dry cleaning tests. Immersion, spray, spot and hydrostatic methods may be used. Shower-resistant, rain-resistant and waterproof factors are interpreted from the results of the testing.

- **water resistant**

  fabric treated chemically to resist water or it may be given a 'wax coating treatment' to make it repellent. Not to be confused with water-repellent. The terms, however, are often used interchangeably.

- **waterproof**

  a term applied to fabrics whose pores have been closed and therefore will not allow water or air to pass through them.

- **weave**

  the process of forming a fabric on a loom by interlacing the warp (lengthwise yarns) and the filling (crosswise yarns) with each other. Filling is fed into the goods from cones, filling bobbins or quills which carry the filling picks through the shed of the loom. Filling may also be inserted into the material without the use of a shuttle, as in the sense, as in the case of a shuttle less loom. The three basic weaves are Plain, Twill and Satin. All other weaves, no matter how intricate, employ one or more of these basic weaves in their composition. There are many variations on the basic principles that make different types of fabric surfaces and fabric strengths.

- **weave constructions**

  1. plain, Tabby, Taffeta: One warp over and one warp under the filling throughout cloth construction.
  2. twill: Diagonal lines on face of cloth-to the right, to the left, or to right and left in same cloth, which gives a broken twill.
  3. satin: Smooth, shiny surface caused by floats of warp over the filling or vice versa. Diagonal lines on the face of the cloth not readily observed by the naked eye, only distinct when seen through a pick glass.

4. basket: Two or more warp ends over and two or more warp ends under in parallel arrangement, interlacing with the filling yarn.
5. rib: Made by cords that run in the warp or filling direction. The corded yarn is covered up by the tight interlacing of the system that shows on the face and back of the cloth.
6. Piqué: Cloth has a corded effect, usually in the warp, but may have cord in filling or both ways. Cords usually held in place by a few ends of plain weave construction.
7. Double Cloth Or Ply Cloth: Two cloths woven together and held in place by binder warp or filling, not pasted, back is usually different from face of cloth.
8. Backed Cloth: Cloth of one warp and two fillings or two warps and one filling. The face is much more presentable than back of cloth. Back is usually dull in appearance. No binders.
9. Pile: Extra yarns from the pile on the face of the cloth. Pile effect may be cut or uncut on the surface. Basic constructions hold cloth in place. Warp or filling yarns may be cut.
10. Jacquard Construction: Pattern or design is woven into cloth reproduction of persons, places or objects, wide range of beautiful designs, used in silks, brocades, etc.
11. Leno, Doup: Warp yarns are paired and half-twisted or fully-twisted about one another, porosity high in some of the cloth designs, two sets of harnesses used- standard and skeleton to cause yarn twisting.
12. Lappet, Swivel, Clipspot: Dots or small figures are woven or embroidered into cloth by use of an extra filling in the case of swivel weave and by extra warp in the lappet weave. Effects are based on plain weave background.

■ **weaving**

the process of producing fabric by interlacing warp and weft yarns.

■ **web**

a sheet of fibres produced by a carding machine (carded web) or combing machine (combed web).

■ **web beam**

another term for the cloth beam.

■ **webbing**

closely woven, stout, strong narrow cotton fabric that has many uses. It ranges in width from less than one inch up to six or more inches. Other fibres are now used to make webbing and there is great demand for all types of elastic webbing. Woven on narrow fabric looms, so-called, as many as 144 pieces of webbing may be woven at one time in a two or three tier arrangement of the warps and shuttles or battens. Uses are for belting of many types, brake-lining, carpet webbing, fan belts, harnesses, surcingles, suspenders and trunk webbing.

Textile

webby wool
a thin fleece with poor staple formation and a large number of cross fibres.

weft
the set of threads which crosses the warp at right angles. Depending on the use, there may be more than one weft used ground weft, pattern weft, etc.

weft knit
a type of knitted fabric in which yarns are formed into stitches in width-wise manner. Common examples of weft knits are circular knits and flat knits.

weft knitting
a method of making a knitted fabric in which the loops made by each weft thread are formed substantially across the width of the fabric. Weft knitting is characterised by the fact that each weft thread is fed more or less at right angles to the direction in which the fabric is produced.

weft-face fabrics
any fabric in which the warp is completely covered with weft.

weight of cloth
there are three ways by which fabric is sold.
1. Ounces per linear yard a 14-ounce covert top coating, a 22-ounce melton over coating.
2. Yards to the pound a 3.60 airplane cloth, a 4.00 filling sateen.
3. Ounces per square yard 3.75 acetate satin, a 6.00 nylon, organdy.

weighted silk
sometimes metallic salts are used in the dyeing and finishing of silk to increase the weight and draping quality-and thus to make it look more expensive. Over-weighting causes deterioration of the fabric.

weighter
any of a great number of agents used to add body, weight or firmness to fabrics. It may be a gum, clay, starch, filler, wax, polyvinyl acetate solution, vinyl acrylic polymer, chloride, salt, sulphide, et al.

weighting
a process by which the weight of a fabric is increased by impregnating it with mineral salts, starch or other materials. This process was often used on yardage that was sold by the pound (e.g. silk).

welt
1. a strip of material seamed to a pocket opening as a finishing as will as a strengthening device.
2. a raised or swelled lap or seam.
3. a covered ornamental strip sewed on a border or along a seam.
4. in knitting, it is flat-knitted separately and then joined to the fabric by looping or hand knitting, as the heel to the stocking.
5. a ribbed piece of knit goods used in forming the end of a sleeve or sock to prevent rolling or ravelling.

wet spinning
in the wet spinning process, the polymer solution (also known as 'dope') is spun into a spin bath containing a liquid chosen for its

■ **wet spun**
a fibre or filament produced by the wet spinning process.

■ **wether**
a male sheep or goat castrated before sexual maturity. Because it is (the operative phrase) is not caught up breeding, all of it is nutrition goes into the fibre thus producing a better quality fibre. Fleeces from wethers cannot be entered into most wool shows but are worth watching for.

■ **wether wool**
any fleece clipped after the first shearing is called wether wool. This wool is usually taken from sheep older than fourteen months and these fleece contains much soil and dust.

■ **wet-laid**
a web of fibres or non-woven fabric produced by depositing an aqueous slurry of fibres on to an endless belt (as in paper making).

■ **wetlaying**
the stage of a production route for making non-woven in which a web of fibres is produced by depositing an aqueous slurry of fibres on to an endless belt (as in paper making).

■ **wet-spun flax**
the process of spinning line flax where the fibres are smoothed by moistening them as they are spun. Traditionally, spinners licked their fingers and drank a lot of beer.

ability to extract the solvent from the dope.

More prosaically, you can get little wooden buckets to hold the water and can hang by a leather thong from the wheel. Or even a bowl of water. You should coat the inside part of the bobbin with paraffin or some other sealant if you don't want to damage your bobbin.

■ **wheel bearings**
the leather, brass, or plastic part of the wheel that holds the 'Axle'.

■ **whipcord**
a steep twill fabric of the gabardine-cavalry twill group which has a very pronounce twill or diagonal on the face of the goods. Of compact texture, the fabric finds use in dress woollens and worsteds, cotton uniform cloth, bathing trunks, livery cloth, public utility uniforms, suitings, topcoats and many types of uniforms used in many areas.

■ **white goods**
a very broad term which implies any goods bleached and finished in the white condition. Some of the cotton white goods are muslin, cambric, dimity, lawn, long cloth, organdy, voile and so on. Tub or washable silks are sometimes classed as with goods, as well as some of the lightweight crepe or sheer woollen or worsted dress goods materials.

■ **white on white**
some fabrics, such as men's shirting of broadcloth, poplin, madras, etc. are made on a dobby or Jacquard loom so that white motifs will appear on a white background. The

madras shirting in this category would have the usual stripe effect with the 'two-tone' white pattern set between these coloured stripes. Some dress goods for summer wear also have the effect.

■ **whorl**

small round grooved bits, on flyer or bobbin to take drive band. The term is also with weaving for the pulley in a draw loom.

■ **wick ability**

the ability of a fabric to transfer liquids, usually perspiration, along its fibres and away from the wearer's skin by capillary action.

■ **wicking**

the passage of fluids along or through a textile material.

■ **wide wales**

materials made by the various degree twill weaves-15, 20, 27, 45, 63, 70 and 75 degrees. To make these effective in cloth material has a diagonal appearance on its face. Wales are distinct in windbreakers, mackinaw cloth, tablecloths of the fancy type and novelty fabrics.

■ **width**

the horizontal measurement on a piece of goods-the breadth, crosswise measurement or filling direction. Textile fabrics are divided into widths, in spans of nine inches. Thus, a ¾ goods would be 27 inches wide; a ⁶/₄ goods, 54 inches wide.

■ **winding**

this spinning term refers to winding the finished yarn onto a bobbin and secured to prevent unravelling.

■ **windle**

a reel or swift.

■ **windproof**

the ability of a fabric or membrane to block the passage of external air through it. In cold climatic conditions, windproof garments help to keep the wearer warm.

■ **wind-resistant**

a limited form of wind-proofing.

■ **wiry wool**

wool that is in-elastic and has poor spinning capacity. It is usually straight and may be the result of poor breeding.

■ **woad**

a glue dye produced from Isatis tinctoria. Not as strong as Indigo.

■ **wool**

strictly speaking, the fibres that grow on the sheep fleece. Wool 'means the fibre from the fleece of the sheep or lamb or the hair of the Angora or Cashmere goat (and may include the so-called specialty fibres from the hair of the camel, alpaca, llama and vicuña) which has never been reclaimed from any woven or felted wool product.

■ **wool classer**

the person who sorts the 'Wool Clip' into the appropriate 'Grades'.

■ **wool classification**

there are about forty breeds of sheep today and counting the cross breeds the total is around 210 distinct grades and types. The classifications follow:

1. Class One: Wools of the merino-breed type, staple length is from one inch up to five inches. Included are Ohio Merino, Silesian of Austria, Saxony of Germany, Rambouillet of France, Australian, South American, South African, New Zealand, plus small amounts of these sheep in Denmark, Italy, Spain and Sweden.

2. Class Two: the better types of the so-called 'hardy types' which originated in England, Scotland, Ireland and Wales and are now raised throughout the world. Staple fibre ranges from two inches to around eight inches. In this class are Bampton, Berkshire, Blackface of Scotland, Cornish, Cornwall, Devonshire, Dorset, Canadian wools, Hampshire, Hereford, Exmoor, Kent, Norfolk, Shropshire etc.

3. Class Three: The so-called 'hardy types' which originated in the British Isles and are now raised throughout the world. These types are inferior to those in Class Two in all respects but provide good to excellent service in apparel and garments made from them. Staple ranges from four to twelve or more inches in length. Known as Lustre Wools, including in this class are Leicester from Leicester County, Cotwold from Gloucester County and so forth.

4. Class Four Wools: Those sheep in wool grading that cannot be classed in one of the first three groups, the result of mixed breeding, the fibre is irregular and ranges from one inch to sixteen or more inches in staple length. The sheep are known as half-breed, semi-lustre or demi-lustre sheep. The wool is used chiefly for making carpets and rugs and low-priced clothing usually for boys and girls.

5. Class Five: This group is not truly a sheep classification but the animals from which the fibres are obtained are akin to those of the wool fibre and are therefore listed. Included are Arabian, Bokharan, Persian lamb and comparable stock raised in many parts of the world.

■ **wool clip**

the total yield of wool short during one season from the sheep of a particular region.

■ **wool combs**

a variety of combs used to produce fibres for worsted spinning. The single-pitch and 2-pitch combs are what you call the peasant combs and are great for producing semi-worsted yarns. English combs are the multi-pitch combs (4 to 5 pitch) used to prepare a true worsted prep.

■ **wool in the grease**

wool in its natural condition as it is shorn from the sheep.

■ **wool roller**

the person in a shearing shed who skirts the fleece, then rolls it. The fleece is then classed.

■ **wool-dyed**

a term applied to yarns where the fibres were dyed prior to spinning, either in the loose fibres or as top or roving..

■ woollen

yarns made from shorter fibres of 1 to 3 inches, which stick out in all directions, giving the yarn its characteristic fuzziness. They are often singles yarns and are thicker and more loosely twisted than worsted yarns. Fabrics made from woollen yarns are warm and fuzzy, such as flannels, tweeds and sweaters.

■ woollen count

the woollen count is based on 1,600 yards of yarn per pound. With woollen yarns, this is called a 'run', so a '2 run' would refer to a 3200 yards.

■ woollen spinning system

in this system, fibre is carded two or three times but not combed and goes directly from cards to the spinning process. Generally wool used for this system are shorter, have more crimp and better felting qualities. With this system it is possible to use wools of different types, lengths and character together in blends.

■ worsted

worsted refers to two different processes which are combined to produce a smooth, clean yarn. Originally it referred to a woollen yarn manufactured in Worstead, Norfolk, England. It now refers to yarn (and fabric) made of long fibres, combed and tightly twisted in spinning. Fabrics made from worsted yarns are smooth and cool to wear.

■ worsted count

the worsted count also expresses the number of hanks required to make a pound of yarn. A hank of worsted wool is equal to 560 yards. So 1 wc = 560 yards of cotton, the coarsest worsted yarn. Worsted sizes are expressed the reverse of cotton sizes. A two-ply number 6 worsted yarn would be expressed as $2/6$ wc and would yield 1680 yards per pound. You can covert worsted count to cotton count by multiplying the cc by 1.5, or wc = cc x 1.5.

■ worsted fabric

a tightly woven fabric made by using only long staple, combed wool or wool-blend yarns. The fabric has a hard, smooth surface. Gabardine is an example of a worsted fabric. A common end use is men's tailored suits.

■ worsted spinning system

a system of yarn production designed for medium or longer wools and other fibres. The suitable fibre lengths vary from 2.5 to 7 inches. The process includes, opening, blending, cleaning, carding, followed by combing, drawing and spinning. These yarns are compact, smooth and more even and stronger than similar yarns spun using the woollen system.

■ worsteds

a wide range of fabrics are made from worsted yarn and are compactly made from smooth, uniform, well-twisted yarns. Little finishing is necessary in these clear surface materials. Plain or fancy weaves are used and the cloth is usually yarn-dyed but piece-dyed fabrics are also popular. Worsted blends are much the vogue today since the major fibres used, nylon and polyester, provide very good service to the consumer. Ideal for summer wear by men and women, some of the fabrics in this fabric family include plain wave worsted, dress goods, gabardine, crepe, serge, tropical etc.

■ **woven fabric**

fabrics composed of two sets of yarns. One set of yarns, the warp, runs along the length of the fabric. The other set of yarns, the fill or weft, is perpendicular to the warp. Woven fabrics are held together by weaving the warp and the fill yarns over and under each other.

■ **woven geotextile**

a geotextile produced by interlacing, usually at right angles, two or more sets of yarns, fibres, filaments, tapes or other elements.

■ **woven-as-drawn-in**

used often with weaving overshot patterns. This is the order of treadling that will give the finished fabric a squared pattern with a diagonal line running through it.

■ **wrap spinning**

a system for manufacturing wrap-spun yarn.

■ **wrap-spun yarn**

a yarn consisting of a core wrapped with a binder.

■ **wrinkle recovery**

some fabrics are able to eliminate wrinkles because of their own resilience. Wool is the best in this group of cloths in wrinkle recovery and thermoplastic manmade fibres and some chemically-treated cottons will recover well. Laboratory tests are now made to determine the amount or degree a fabric will recover from wrinkling.

■ **wuzzing**

a way of removing excess moisture by centrifugal force. Hold onto the end of the skein firmly and spin it around like a helicopter blade.

■ **xanthating**

a process in making rayon. The chemical treatment of cellulose in which carbon disulphide reacts with alkali cellulose crumbs to produce the bright orange cellulose xanthate.

■ **yacht cloth**

stoutly made, unfinished worsted in blue, white and delicate stripes. Used in yachting circles.

■ **yardage**

any fabric made and sold by the yard.

■ **yarn**

a continuous strand of textile fibres that may be composed of endless filaments or shorter fibres twisted or otherwise held together. It may be made up of vegetable (linen, hemp, jute, sisal, ramie, cotton), animal (wool, mohair, silk), or artificial fibres (gold, silver and other metals rayons, nylon, Orlon). Yarns are utilised in making fabric. Yarn is characterised by its composition, its thickness (or grist or count), number of strands (or plies), direction and degree of twist and the colour.

- **yarn dyeing**

yarn that has been dyed prior to the weaving of the goods, follows spinning of the yarn. May be done in either total immersion or partial immersion of the yarn.

- **yarn-dyed**

a term used to describe a design or fabric which is constructed from and coloured by means of, pre-dyed yarns.

- **yearling**

a sheep or goat that is 12-18 months of age.

- **yellowing**

a white fleece can yellow for a variety of reasons. Yellow stains can be caused by urine or faeces and by bacteria or fungus. Alkali and light can also cause yellowing in wool. Most stains cannot be removed by washing. Bleaching will probably damage the fleece.

- **yield**

the amount of scoured wool obtained from a definite amount of grease wool. The amount of usable fibre after the processes of washing, drying and removing guard hairs. A 'high yield' fleece would have a low percentage of grease.

- **yolk**

the natural grease and suint covering on the wool fibres of the unscoured fleece and excreted from glands in the sheep's skin. Usually the finer the wool, the more abundant the yolk. Yolk serves to prevent entanglement of the wool fibres and damage during growth of the fleece.

- **zero twist**

sometimes referred to as 'no twist'. The thrower (spinner) may request that viscose yarns be supplied with no twist. This is rarely done, however, usually 1 to 5 turns per inch are given the yarns. Viscose and acetate yarns with from 3 to 5 or 6 turns per inch are normally supplied to the thrower, this twist is known as 'tram twist' or 'filling twist'.

- **zibeline**

used for cloaking, coats and capes in women's wear. The cloth is made from cross-bred yarns and the fabric is strongly coloured. Stripping, sometimes noted in the cloth, work in very well with the construction and appearance of the finished garment. The finish is a highly raised type, lustrous and the nap is long and lies in the one direction. The cloth may or may not be given a soft finish and feel.

■ **z-twist**

if the spirals in the yarn conform in slop to the central portion of the letter Z, then the twist is classed a Z-twist. Formerly known as right-hand or counter-clockwise twist..

Other Titles are Available in
LOTUS ILLUSTRATED DICTIONARIES

1. First English Dictionary **(New)**
2. Astronomy — 95/-
3. Agriculture — 95/-
4. Anthropology — 95/-
5. Archaeology — 95/-
6. Architecture — 95/-
7. Art — 95/-
8. Banking Finance & Accounting — 95/-
9. Bio - Chemistry — 95/-
10. Business Administration — 95/-
11. Bio - Technology — 95/-
12. Biology — 95/-
13. Botany — 95/-
14. Culture — 95/-
15. Chemical Engg. — 95/-
16. Chemistry — 95/-
17. Civil Engineering — 95/-
18. Computer Science — 95/-
19. Commerce — 95/-
20. Cooking & Food — 95/-
21. Ecology — 95/-
22. Economics — 95/-
23. Education — 95/-
24. Electrical Engineering — 95/-
25. Electronic & Telecommunication — 95/-
26. Environmental Studies — 95/-
27. Festival — 95/-
28. Geography — 95/-
29. Geology — 95/-
30. Genetic Engineering — 95/-
31. Health & Nutrition — 95/-
32. History — 95/-
33. Import and Export — 95/-
34. Information System Management — 95/-

#	Subject	Price
35.	Internet	95/-
36.	IT	95/-
37.	Inorganic Chemistry	95/-
38.	Law	95/-
39.	Library & Information Science	95/-
40.	Literature	95/-
41.	Management	95/-
42.	Mathematics	95/-
43.	Marketing & Sales	95/-
44.	Mass Communication	95/-
45.	Mechanical Engg.	95/-
46.	Medical	95/-
47.	Music	95/-
48.	Organic Chemistry	95/-
49.	Philosophy	95/-
50.	Physical Education	95/-
51.	Physics	95/-
52.	Psychology	95/-
53.	Science	95/-
54.	Sex	95/-
55.	Sociology	95/-
56.	Sports	95/-
57.	Textile	95/-
58.	Zoology	95/-
59.	Dictionary of Veterinary Sciences	125/-

4263/3, Ansari Road,
Darya Ganj, New Delhi-110 002
Ph.: 32903912, 23280047, 9811594448
E-mail: lotus_press@sify.com
www.lotuspress.co.in